How to Do
Ecology

How to Do
Ecology

A Concise Handbook

THIRD EDITION

Richard Karban
Mikaela Huntzinger
Ian S. Pearse

Princeton University Press

Princeton and Oxford

Published by Princeton University Press
41 William Street, Princeton, New Jersey 08540
99 Banbury Road, Oxford OX2 6JX

press.princeton.edu

All Rights Reserved
ISBN (pbk.) 9780691245751
ISBN (e-book) 9780691250434

Library of Congress Cataloging-in-Publication Data

Names: Karban, Richard, author. | Huntzinger, Mikaela, author. | Pearse, Ian S., author.
Title: How to Do Ecology : A Concise Handbook / Richard Karban, Mikaela Huntzinger, Ian S. Pearse.
Description: Third edition. | Princeton, New Jersey : Princeton University Press, 2023 | Includes bibliographical references and index.
Identifiers: LCCN 2022056533 (print) | LCCN 2022056534 (ebook) | ISBN 9780691245751 (pbk.) | ISBN 9780691250434 (e-book)
Subjects: LCSH: Ecology—Research—Handbooks, manuals, etc. | Ecology—Experiments—Handbooks, manuals, etc. | BISAC: SCIENCE / Life Sciences / Ecology | SCIENCE / Research & Methodology
Classification: LCC QH541.2 .K374 2023 (print) | LCC QH541.2 (ebook) | DDC 577.072—dc23/eng/20221219
LC record available at https://lccn.loc.gov/2022056533
LC ebook record available at https://lccn.loc.gov/2022056534

British Library Cataloging-in-Publication Data is available

Editorial: Sydney Carroll & Sophia Zengierski
Production Editorial: Jaden Young
Jacket/Cover Design: Wanda España
Production: Lauren Reese
Publicity: Matthew Taylor
Copyeditor: Jodi Beder

Jacket image: Denis Shevchuk / Alamy Stock Photo

This book has been composed in ITC New Baskerville

10 9 8 7 6 5 4 3 2 1

Table of Contents

vii   *List of Illustrations*

ix   *List of Boxes*

xi   *Preface to the Third Edition*

xiii   *Introduction: The Aims of This Book*

1   CHAPTER 1.
          Picking a Question

17   CHAPTER 2.
          Picking an Approach

35   CHAPTER 3.
          Testing Ecological Hypotheses

61   CHAPTER 4.
          Using Statistics in Ecology

76   CHAPTER 5.
          **Using Quantitative Observations
          to Explore Patterns**

96   CHAPTER 6.
          **Brainstorming and Other
          Indoor Skills**

115   CHAPTER 7.
             **Working with People and Getting
             a Job in Ecology**

135   CHAPTER 8.
             Communicating What You Find

190   CHAPTER 9.
             Conclusions

193   *Acknowledgments*

195   *References*

207   *Index*

Illustrations

26–27   FIGURE 1. Apparent competition
         between white-tailed deer and caribou

47   FIGURE 2. A scope (scale) diagram
         showing observational, experimental,
         and modeling approaches used to
         characterize changes that occur
         to harbors over decades

51   FIGURE 3. Experimental designs to
         evaluate the effects of predators on
         greenhouse pests

66   FIGURE 4. The relationship between
         the number of stork pairs and the
         human birth rate in 17 European
         countries

78–80   FIGURE 5. A path analysis showing
         various hypotheses about effects of
         logging, roads, fire suppression,
         deciduous trees, and wolves, deer, and
         moose on numbers of caribou in
         northwestern Ontario

88   FIGURE 6. Possible relationships
         between March temperature and bird
         breeding

Boxes

6	Box 1. Advice for three types of ecologists on picking questions
15	Box 2. The importance of research for people who aspire to non-research careers
69	Box 3. Generating alternative hypotheses
101–105	Box 4. Generating ideas
139	Box 5. Steps in the writing process
155–157	Box 6. Journal article checklist
173–175	Box 7. Oral presentation checklist
178–179	Box 8. Poster checklist
186–187	Box 9. Grant proposal checklist

Preface to the Third Edition

This book started out as handouts for Rick's graduate-level field ecology course. At first, it was just a few pages, but each year the stack grew thicker. Nothing more happened for about a decade until Mikaela, in the throes of graduate school herself, suggested that the collection of handouts might be useful to ecologists outside the field course. She added material and her perspective and organized it into a book.

We had no idea who would want to read this thing. However, the first edition taught us that most of our readers are prospective and current grad students in ecology. So in the second edition, we added advice that we hoped would be valuable to students (see Introduction). Also, in the meantime, the world of ecology had changed—for example, more ecologists use surveys and other observational techniques to test hypotheses about nature. So for the second and third editions, Ian joined Rick and Mikaela to add his own expertise.

In the third edition, we have included a completely revised brainstorming "workshop" that you can use in a group or even on your own. In addition, more grad students in ecology are seeking nonacademic careers answering applied ecological questions, so we address more career paths. Finally, the statistical techniques that ecologists

commonly use have changed with the widespread adop-
tion of mixed models in R and with Bayesian approaches;
we have added some of these.

We hope this book makes learning to do ecology more
straightforward for you than it was for us.

Introduction:
The Aims of This Book

Different games have different currencies. In basketball, the currency is buckets; in *Catan*, the currency is settlements, development cards, the longest road, and the largest army. In your high school and undergraduate days, the currency was grades or test scores. Over many years as a student (most of your life?), you probably invested a lot of time in developing the skills you'd need to accumulate good grades and high test scores.

When you enter grad school, the currency changes, often without much acknowledgment. Nobody cares about your classes, grades, or scores (there are a few exceptions, but not many). It doesn't matter anymore if you get an A+ instead of a B+. The currency in grad school and on the ecology job market is research publications and, in some cases, grants. One of the aims of this book is to make you aware of this shift so that you can conduct and publish research to earn this new currency. Ultimately, you'll probably find it more satisfying to develop research skills than to rack up grades and scores anyway.

To succeed in grad school, prioritize the activities that will result in publications. When allocating your work time, think carefully about activities that won't result in pubs; you may be drawn to them because they are easier or more familiar, not because they will get you to the finish line. The only thing that will earn the currency you need is writing up your chapters and papers. It is also important to take care of your mental health, so we don't mean that you should put off exercising or socializing with friends and family. We mean that you should prioritize designing, conducting, and publishing your research before you focus on other skills, like teaching your own course, learning a technique that's not directly related to your study, or building your social media presence.

Both undergrad and grad school require classes in ecological principles and theory; this emphasis allows you to examine the studies that have shaped our discipline. However, not many classes explicitly address how to ask ecological questions or conduct ecological studies. This book attempts to help you do just that.

In this book, we consider different ecological approaches (manipulative experiments, quantitative observations, and models) to test hypotheses, and we discuss their strengths, weaknesses, and utility. We present some rules of thumb for setting up manipulative and quantitative observational studies as well as analyzing and interpreting results. We offer suggestions about generating creative ideas, organizing your field season, working with other people, applying for jobs in and outside of academia, communicating your research, and other things ecologists might need to know. Not everything we say here will apply to you, but we hope this book offers you some topics to think about and discuss with your mentors and peers.

How to Do
Ecology

CHAPTER 1
_____

Picking a Question

To contribute to progress in ecology, you need to generate new knowledge. This is quite different than memorizing what is already known. As a result, perhaps the most critical step in doing field biology is picking a research question. Tragically, it's the thing that you are expected to do first, when you have the least experience. For example, it is helpful if your application essay for grad school appears to be focused on a particular set of questions that matches a professor's interests. However, at this stage in most students' careers, many topics sound equally interesting, so forcing yourself to focus in this way is daunting or even painful.

The gold standard: Novel, general, feasible, exciting, and not perfect

Your research question should be as novel as possible. All projects have to be original to some extent. We all like to hear new stories and new ideas, and ecologists place a large premium on novelty. If you are asking the same question that has been answered in other systems (that is, with similar organisms in analogous environments), it behooves you to think about what you can do to set your study apart from the others. That said, if you are trying to start a project and haven't yet thought of a novel question, a useful way to begin may be to repeat a study that captured your attention and imagination, but with a different organism or system. Your repeat can be quick and dirty (not too many reps, not too long term), just enough to inspire an exciting new direction for your own novel question.

Policy makers are much less concerned with novelty than academics are. If you are funded by an agency to answer a specific policy question, you will need to balance your academic colleagues' expectations of novelty and your funding source's demands to answer the specific question for which you are funded. Your first priority should be to generate relevant data for your funders; however, if possible, ask additional, complementary questions in your study system that can lead to publishable research.

In addition to novelty, ecologists like generality. That is, we get more excited about broad or theoretical questions than about specific, narrow ones. It is possible to ask a question that is too general, especially if you are building a model; in that case, ask yourself if your answer will reflect reality for at least one actual species or habitat. It is more common for students to find themselves answering an overly specific question that may be considered important by only a very small community. If possible, ask a question that has the potential to matter beyond your study system or organism. If your question is very specific, ask whether you can generalize from your results. For example, you may find yourself answering a question about managing a specific fishery, restoring a particular plant species, and so on if your funding comes from an applied source. It may not be possible to couch your question in more general terms. Instead, you may be able to ask a complementary, more conceptual question as well. For example, let's say you have been funded to determine which animals visit a particular endangered night-blooming flower. More general (and interesting) questions might be which of those animals successfully fertilize that species and what characteristics of the flower and/or the visitors make them effective pollinators. Are the traits that you identify shared by other night-blooming species? The answers to these latter questions will be compelling to a broader audience.

A relatively small question can catalyze a general question. By "small," we mean specific to your study system and with relatively little replication. Small questions will often generate more excitement for you than bigger ones because their more modest goals can be achieved with relatively few data, and much more quickly. Imagine that you want to study rates of predation on Canada goose eggs. These eggs may be difficult to find and highly seasonal. So, you could conduct a small pilot experiment with three cartons of chicken eggs from the grocery store. Your pilot study will not give you definitive answers about goose eggs but will likely provide useful insights about how to conduct that experiment. You don't need to invest an entire season on a pilot study. Do simple analyses of your data early and often. If results from the pilot study turn out as expected, they can provide a foundation for a bigger project. If the results are unexpected, they can serve as a springboard for a novel working hypothesis. Almost all of our long-term projects had their beginnings as pilot "dabbles."

A third component of a good research question is feasibility. Many of the factors associated with failure or success in field projects are beyond your control. Nonetheless, you should ask whether your ideas are feasible—are you likely to get an answer to the questions you pose? Do you have the resources and knowledge to complete the project? Your armchair answers to these questions won't necessarily be on target, but they can help you anticipate and plan for potential problems. Don't talk yourself out of doing an exciting project just because it seems challenging—think about ways to make it work.

Since most field projects don't work, try several pilot studies and follow the leads that seem the most promising. If you know that you want to ask a particular question, try it out on several systems at the same time. You'll soon get a sense that the logistics in some systems are easier than in others and that the biological details make some systems more amenable to answering particular questions. It is a lucky coincidence that Gregor Mendel worked on peas since they are particularly well suited to elucidating the particulate nature of inheritance. Other people had attempted to ask similar questions but were less fortunate in the systems that they chose to investigate. Don't get discouraged about the ones that don't work. Successful people never tell you about the many projects (journal submissions, job applications) they didn't pull off. You should feel fortunate if two out of seven projects work well.

An essential ingredient of a good project is that you feel excited about it. The people who are the most successful over the long haul are those who work the hardest. No matter how disciplined you are, working hard is much easier if it doesn't feel like work but rather something that you are passionate about. As Kong Fuzi (previously known as Confucius in the West) is supposed to have said, "If you have a job you love, you will never have to work a day in your life." Dehua Wang, a professor of zoology at Shandong University, has told us that a better translation is something like: "Those who know are not as good as those who want to know, and those who want to know are not as good as those who are driven to know." The message here is to pick a project that is intellectually stimulating, specifically *to you*. Figure out what you are really driven to know. You are the one who must be excited enough about it to do the boring grunt work that all field projects involve. You will feel much more inclined to stay out there in the pouring rain, through all the mind-numbing repetitions that are required to get a large enough sample size, if you have a burning interest in your question and your system.

So, you're looking for questions that are specific yet general, and novel yet relevant to your interests. You could fret over this for years. Don't agonize over the perfect study before you are willing to begin (see box 1). One of the most unsuccessful personality traits in this business is perfectionism. Field studies are never going to be perfect. For example, don't get stuck thinking that you need to read more before you can do anything else. Reading broadly is great,

Box 1. *Advice for three types of ecologists on picking questions*

There are three kinds of ecologists:

· The perfectionist who waits for a transformational idea before starting,
· The jackrabbit who has a lot of energy and wants to get started before thinking through their goals and their study, and
· The Goldilocks who is just right, someplace in between.

If you are a perfectionist who can't get started because you haven't thought of the perfect question, we suggest you just go out there and do it. The experience and insight (not to mention publications) that you'll get by doing an imperfect study will help you improve in the future. If you are a jackrabbit and find yourself starting a million projects, our advice is to take a moment and ask which of these questions is most likely to advance the field and, even more importantly, inspire enduring passion in you. And if you are a Goldilocks who has it just right, maybe post that you are humbled by your own success.

but you will learn more by watching, tweaking, and thinking about your system. Talk to many people about your ideas—your major professor, peers, family, and so on. It is not realistic to expect yourself to sit at your desk and conjure up the study that will revolutionize the field. Revolutionary questions don't get asked in a vacuum; they evolve. This is one reason repeating a past study (see above) can be a useful springboard. We often start asking one question, hit a few brick walls, and get exposed to some ideas or observations that we hadn't previously considered; and then pretty soon we're asking very different questions that are better than our initial, naïve ones. Most projects don't progress as we originally conceived them.

How to pick a project

There are two approaches to picking a project: starting with the question or starting with the organism or system. The difference between these two is actually smaller than it sounds, because you generally have to bounce between both concerns in order to come out the other side with a good project. So regardless of which one you start with, you need to make sure that you are satisfying a list of criteria related to both.

Starting with a question

Many successful studies start with a question. You may be interested in a particular kind of interaction or pattern for its own sake or because of its potential consequences. For example, you may be excited by the hypothesis that more diverse ecological systems are intrinsically more stable. Maybe you're interested in this hypothesized relationship because if it is generally true, it could provide a sound rationale for conserving diversity, and if it is not generally true, ecologists should not use it as a basis for conservation policy. Since many studies have considered this question, you should think about what's at the bottom of the hypothesized link between biodiversity and stability. Have previous studies addressed these key elements? Are there novel aspects to this question that haven't been addressed yet? Are there assumptions that scientists take for granted but have never tested? Even questions that have been addressed by many researchers may still have components that have yet to be asked.

If you start by asking a question, you will need to find a suitable system (that is, interacting species and their surroundings) to answer it. The system should be conveniently located. For example, if you don't have money for travel, choose a system close to home, and if you don't like to hike, choose plots close to the road. Your study organisms or processes should be common enough for you to get enough replication. Ideally, your sites should be protected from vandalism by curious people and animals (or it should be possible for you to minimize these risks). Your system should be amenable to the manipulations that you would like to do and the observations you would like to make. You can get help finding systems by seeing what similar studies in the literature have used, by asking around, or by looking at what's available at field stations or other protected sites. The appropriate system will depend upon the specific questions that you want to ask. If your question requires you to know how your treatments affect fitness, you will want to find a short-lived species rather than a charismatic but long-lived species. If your hypothesis relies upon a long history of coevolution, you should probably consider native systems rather than species that have been recently introduced. (Incidentally, there is a widespread chauvinism about working in pristine ecosystems. The unspoken assumption seems to be that the only places where we can still learn about nature are those that have not been altered by human intervention. Certainly, less disturbed places are inspiring and fun, but they also represent a very small fraction of the earth's ecosystems. There are still plenty of big questions about how nature works that can be asked in your own backyard regardless of where you live—we can attest to this, having lived in some truly uninspiring places.)

Be careful that you aren't shoehorning a system to fit your pet hypothesis. If you start with a question, look around for the right system for that question and be willing to modify your question as necessary to go where the natural history of your chosen system takes you. You cannot make your organisms have a different natural history, so you must be willing to accept and work with what you encounter.

It is also possible that you will be handed a question, particularly if you are a master's student. This has its benefits and drawbacks. You don't have to come up with your own hypothesis; on the other hand, you may not feel as much ownership of or excitement about your research. Also, you may not learn how to pose a good question, which is one of the most important skills you can take from grad school. If your major professor agrees, you may be able to add your own question as well.

Starting with a system

If you start with an organism or a system because of your interests, your funding, your major professor, whatever, you may find yourself in search of a question. Try skimming the literature broadly to get a sense of the kinds of questions that are exciting and interesting to you (see chapter 6 for reading strategies); you may be able to apply those questions to your organism or system. Often an organism becomes a model for one suite of questions but has not been explored for others. For example, the genetics of *Drosophila* and *Arabidopsis* are well studied in the lab, but their ecologies are poorly known in the field. Similarly, sometimes a system becomes popular for one type of question, but no one has asked the question you're interested in. (In those systems, previous studies may also offer you valuable background natural history.) For example, nectar was long assumed simply to be sugar water that attracted and rewarded pollinators. Tadashi Fukami and his lab knew that nectar contained diverse microbes and used microbes in nectar to develop novel models about community assembly (Peay et al. 2012). As a postdoc in this lab, Rachel Vannette then used the same system to ask questions about how these nectar microbes affected plant-pollinator interactions (Vannette et al. 2013).

Let's say you don't have an organism *or* a system. Try going to a natural area and spending a few days just looking at what's there. As you poke around, generate a list of systems and patterns. For those that interest you most, gather quick-and-dirty quantitative and qualitative data. For example, you might observe that snails are at a particular density at your study site. Next, ask whether there is natural variation in this measurement. Do some microhabitats have more snails than others? Is there natural variation associated with behavioral traits? For example, are the snails in some spots active but those in other spots aestivating? Is there variation between individuals? Are the snails in some microenvironments bigger than those in others? And so on. Once you have quantified these patterns, ask more about them. What mechanisms could cause the patterns that you observe? What consequences might the patterns have for individuals and for other organisms? Traits are shaped by natural selection, so ask questions about survival and reproductive output when you can. Using this technique during Rick's field class, students invariably come up with more good questions than they can pursue.

Mechanisms and consequences

Even if a pattern you observe in your scouting has been described before, it may still form the basis of many novel projects. If it is an important and general pattern, other people have probably noticed it too. However, it is less likely that the ecological mechanisms that cause the pattern have been evaluated. Understanding ecological mechanisms not only provides insight into how a process works, but also can tell us about its effects and where we would predict it to occur. Elucidating the mechanisms of a well-known pattern is likely to be a valuable contribution. Generate a list of potential mechanisms and then devise ways to collect evidence to evaluate the strength of each.

It is also less likely that the consequences of the pattern have been described. Does the pattern affect the fitness of the organisms involved and under what conditions? Does it affect their population dynamics? Does it affect the behaviors of organisms in the system? Answering any one of these questions is plenty for a dissertation.

Don't assume that questions have been answered just because they seem obvious. For example, thousands of studies have documented predation by birds on phytophagous insects, but the effects of that predation on herbivory rates and plant fitness went relatively unexplored for decades (Marquis and Whelan 1995, Mooney et al. 2010). Also, although periodical cicadas are the most abundant herbivores of eastern deciduous forests of North America, their interactions with their host plants and the rest of the community are largely unexplored. Several hundred years into studies of the natural history of periodical cicadas, Louie Yang (2004) found that the pulses of dead cicada adults stimulated soil microbes and altered plant communities.

Scores of ecologists have observed that individual organisms vary from one to another, but most have dismissed that variation as noise and focused only on the average tendencies. This reflects a general tendency to look for overall trends and to disregard variation. Consequently, variation among individuals is a promising source of questions. For example, recent work looking at individual variation among conspecific animals has found that it can be interesting and important in its own right (Sih et al. 2012). Similarly, the extent of spatial and temporal variation in plant traits has been found to be as impactful to herbivores as the mean values of the traits themselves (Wetzel et al. 2016). In both of these examples, innovative advances have been made by considering variation around the mean values of traits even though the trait means themselves had been well studied. In short, there are still many interesting unanswered questions even in well-known systems.

Telling a complete story

Your ultimate goal will be to tell one complete story, which will be more compelling and satisfying than a haphazard assortment of loosely related pieces. So, once you have selected a question and collected some preliminary data, think about how to develop your story as fully as possible. Keep in mind that no story is ever truly comprehensive. Here are some additional questions that could make your study more complete.

1. Think about whether the phenomenon you are studying applies generally. For instance, you may want to repeat your studies that had interesting results at other field sites or with other species.
2. If possible, work at levels both upstream (mechanisms) and downstream (consequences) of the level of your pattern. What ecological mechanisms could generate the pattern that you observe? What other organisms or processes could the pattern affect?
3. Explore whether your phenomenon operates at realistic spatial and temporal scales (see chapter 3). For instance, if you conducted an experiment at a small spatial scale, do your results apply at the larger scales where the organisms actually live?
4. Consider alternative hypotheses that could produce the patterns and results you observe (see chapter 4).

The more complete your story is, the more useful and appreciated your work is likely to be. Each of these additional questions can take a lot of time and energy, so don't expect to address them all. Prioritize the questions that flesh out your best story and the questions that you can feasibly answer.

The bigger picture: Your question should reflect your goals

The question that you pick should reflect your goals as a biologist. It's worth figuring out what your short-, mid-, and long-term goals are and then making a plan to help you achieve them. If you are a new grad student, your short-term goal might be nothing more than to succeed in grad school. Make sure you don't focus on gaining what you believe will be marketable skills at the expense of doing something you are passionate about. It's important to look farther down the road even as you're beginning. Try to pose a question that is deeply interesting to you.

A common mid-term goal is getting your first job. For most jobs (those at research universities, small liberal arts colleges, federal agencies, and nonprofit organizations), search committees want to see a strong record of research and publication even if you won't be expected to do research or publish a lot on the job. Box 2 presents a justification for this bias. Search committees want to know that you are capable of advancing the field and communicating effectively. (They may also want to see other qualifications and experiences, such as teaching, grant-writing, or outreach; see chapter 7.)

Box 2. *The importance of research for people
who aspire to non-research careers*

Even if a career in research is not one of your long-term
goals, it is still worth throwing yourself into the world of
research while you work on your degree. The process
of doing research will teach you things that are hard to
absorb and integrate in any other way.

· Testing your own hypotheses helps you understand how
  individual biases, preconceptions, and points of view shape
  the ecological information that appears in textbooks.
· Over time, working on independent research helps you
  to incorporate scientific reasoning into your everyday
  thinking, which allows you to analyze reports and articles
  critically and to teach the information to others more
  effectively.
· Even if you are already a strong communicator, writing up
  your results will teach you how to write more efficiently,
  concisely, and clearly.
· Analyzing your own data is a much more compelling way
  to absorb important abstract ideas and analytic tools
  than trying to learn them from homework sets.

These and other insights and skills are virtually impossible to gain solely through reading; instead, you are more
likely to learn by truly immersing yourself in your research. And besides, it's fun.

Your mid-term plan will probably revolve around a larger
suite of questions than your short-term plan. For example,
your plan might include solving a problem in restoration,
such as how to return a particular piece of real estate to some
level of ecological functioning. A more conceptual mid-term

goal might involve making people rethink the interactions that drive the abundance or distribution of a taxon.

Long-term goals are harder to formulate but are at least as important. (If you don't believe this, talk to some burnt-out researchers late in their careers. Some people never stopped to figure out what they really valued and wanted to accomplish for themselves. Thinking through your big-picture, long-term goals makes doing the work more enjoyable.) Some long-term goals that you might want to try out include attempting to influence how you and others think about or practice a subdiscipline of biology, how to manage a crop, or how to recognize and mitigate some effects of climate change.

Your long-term goals should suit you and not necessarily your major professor (who may consider nonacademic goals a waste of time), and not necessarily your parents (who may try to convince you that a conceptual thesis will leave you unemployable). While you shouldn't let uncertainty about your long-term interests slow down your research progress, having long-terms goals in mind can provide a yardstick with which to evaluate your choice of project.

In summary, allow your organisms to direct your questions. Many discoveries in science are unplanned. While you are answering one question, you are likely to see things that you haven't imagined. There is some chance that nobody else has seen them either. Rather than trying to force your organisms to answer your questions, allow them to suggest new ones to you. Read broadly so that you recognize that something is novel when you stumble upon it. Above all, be opportunistic!

CHAPTER 2

Picking an Approach

So, the right questions are novel but feasible. They are grounded in an organism or system, and they address your research goals. What you can learn about ecology depends not only on the questions you ask but also on the approaches you use to answer those questions.

Different ways to do ecology

Ecologists use three main approaches to understand phenomena: quantitative observations, manipulative experiments, and model building. These approaches aren't mutually exclusive, and each has something to offer. Deciding on an approach may sound like a bunch of philosophical nonsense, but each one constrains the kinds of results and insights we get.

Quantitative observations

Observations of patterns in natural systems are essential, as they provide us with the players (organisms and processes) that may be important. Observations allow us to generate hypotheses and to test models.

Natural history (that is, observing and quantifying organisms and interactions) used to be the mainstay in ecology,

but it started to go out of style in the 1960s. Current training in ecology has become less and less based on a background in natural history (Ricklefs 2012, Anderson 2017).
Undergraduate education requires fewer hours of labs
than it did in the past for economic rather than educational
reasons: labs are expensive and time-consuming to teach.
Traditional courses in the "-ologies" (entomology, ornithol-
ogy, herpetology, etc.) are endangered. Graduate students
are pressured to get started on thesis projects before they've
spent time poking around in real ecological systems. What's
more, professors are most "successful" when they adminis-
ter research while their students or employees do the hands-
on work. Many professors write grants to fund other people
to do the work with real organisms in the name of writing
progress reports and the next grants. As a result, the inspiration for our studies often comes from the literature, the
computer screen, or our major professor, not from something new we noticed in the field. We spend a lot of time
refining an old observation that everyone already believes
is important. This cycle has the danger of making ecology
conservative and unexciting.

It is clear to us that the discipline of ecology would be
improved if we were encouraged to learn more about nature by observing it first and manipulating and modeling
it second. Observations provide the insights that make for
good experiments and models (see sections below). As
experimenters, we can only measure a limited number of
factors; the factors we choose to include determine which
factors we will conclude are important. For instance, if we
test the hypothesis that competition affects community

structure, we are more likely to learn something about competition and less likely to learn something about some other factor we didn't think to manipulate, such as mutualisms. Observe your organism or system with as few assumptions as possible, and let your system suggest what factors to measure.

Good intuition is the first requirement for designing meaningful studies. The best way to develop that intuition is by observing organisms in the field. Sadly, few of us "have the time" to just observe nature. Graduate committees and tenure reviewers are not likely to recommend investing precious time in this way. However, observations are absolutely essential for you to generate working hypotheses that are grounded in reality. So, carve out some time to get to know your organisms. If you are too busy with classes and other responsibilities, then reserve two days before you start your experiments to observe your system with no manipulations (or preconceived notions). It's often fun to do this with a lab mate or colleague. The opposite can work well too: consider spending a whole day with no other people or distractions around, just looking at your system.

After you have set up your manipulations, continue to monitor the natural variation in your system. This will help you interpret your results and plan better studies for the next season. For example, Mikaela's initial plan for her first research project involved examining the role of fire on butterfly assemblages on forested hillsides. Poking around during her first season revealed that most butterflies were using riparian areas, a habitat that fire ecologists had largely ignored. This led to a second experiment the following year that was far more informative than the original experiment she had planned (Huntzinger 2003).

Record your observations in a real or virtual field notebook. It's important to do this because they are difficult to remember and you don't know which observations will be relevant. Also jot down ideas that you have in the field about your study organism, other organisms that it may interact with, or even general ideas about how ecology works. Decide on a standardized method for recording observations. Science requires a lot of information management, so if you aren't consistent, you may have trouble retrieving information when you need it to flesh out your methods or spark new ideas.

If you don't know where to begin with your observations, try quantifying a pattern. Common ecological patterns include changes in a trait of interest that varies over space or time. This could be anything from a physical trait of individuals (e.g., beak length) to a trait of ecosystems (e.g., primary production or species richness). Ask questions about it. How variable is the trait? Is there a pattern to the variation over space or time? For example, are there large differences in primary production from one place to the next? What factors correlate with the variation that you observe? For example, are differences in primary productivity correlated with differences in diversity? It is often helpful to represent the pattern as a figure with one variable on the x-axis and the other on the y-axis. This representation allows you to get a sense of the pattern—how strong it is and whether the relationship between the two variables appears linear. If you find a pattern, a manipulative experiment can help to determine whether the two variables are causally linked.

In some cases, observations of patterns even replace manipulative experiments as the best way to gain ecological understanding. This is due in part to the unhappy fact that many processes are difficult to manipulate experimentally. Manipulative experiments are frequently conducted on small plots and over short periods of time (Diamond 1986). However, important ecological processes often occur at larger spatial or temporal scales that are difficult or impossible to replicate. For example, catastrophic fires and floods can be important ecological drivers but can't be experimentally imposed at realistic scales. Manipulations involving vertebrate predators are also difficult to achieve with any realism since their home ranges are often larger than the plots available to researchers. Removing predators is often more feasible than adding them, although any density manipulations, such as killing organisms, may be unethical. Manipulating processes that involve endangered species may be not just unethical but even illegal. When experiments are logistically or ethically problematic, observations are often the best approach. Observational tests of hypotheses still require replication and controls to be most informative. We discuss how to analyze observed patterns in chapters 4 and 5.

Although it's tempting to extrapolate results from small-scale experiments to more interesting and realistic processes at larger scales, it is difficult to justify doing so. One partial solution to this dilemma is to observe processes that have occurred over larger spatial and temporal scales and ask whether these observations support results from modeling and small-scale experiments. Such large-scale observations are sometimes termed "natural experiments" since the investigator does not randomly assign and impose the treatments (Diamond 1986). For example, using satellite images to compare numerous sites that had been cleared of brush with those that had not been cleared provided more general information than did a smaller scale manipulative experiment at a single site (Fick et al. 2021).

Long-term data sets expand the temporal scale of any experimental or observational study. It's worth considering linking your work to a long-term survey. For example, daily surveys of amphibians have been collected at one pond, Rainbow Bay in South Carolina, since 1978. This record has been useful for understanding the causes of worldwide amphibian declines (Pechmann et al. 1991), the consequences of anthropogenic climate change (Todd et al. 2011), the importance of connectivity among spatially separated populations (Smith et al. 2019), and other ecological issues.

Despite having a lot to offer, observations are often less valued than manipulative experiments. Quantitative observations can be applied to test hypotheses, but they are poor at establishing causality. For example, we can observe that two species do not co-occur as frequently as we would expect. This may suggest that the two species are competing. In the early 1970s everyone was "observing competition" of this sort because it seemed to make such good theoretical sense. However, the observed lack of co-occurrence could

be caused by the two species independently having different habitat preferences that have nothing to do with current competition. For example, Robert MacArthur's fa-
mous study documented that different warbler species fed
in different parts of the spruce canopy. MacArthur's observations suggested that foraging overlap was responsible
for this distribution, although intensive field experiments
later found that nesting habitat was more likely the cause
(Martin 1993). Quantitative observations alone do not allow
the causes of the pattern to be determined, though methods have been developed to infer causal relationships from
observational data (chapter 5). Observational studies are
often valuable components of publications, so record data
from observations as carefully as those from manipulative
experiments.

Manipulative experiments

A manipulative experiment varies only one component
of the system at a time (or at most a few). The experimenter controls the conditions, termed the "treatment,"
that are imposed on the members of each experimental
group. The treatment group can be compared to a base-
line group, the "control." If the experiment has been set
up properly, any responses can be attributed to the manipulation (treatment). Experiments may include multiple levels of the treatment or multiple manipulations
to compare different treatments simultaneously. This approach is very powerful for establishing causality. Statistical tests can be used to evaluate the likelihood that the
observed effect was caused by chance rather than by the

manipulation. These issues will be considered in much more
detail throughout this book, particularly later in this chapter ("Why ecologists like experiments so much") and in
chapter 3.

Model building

Modeling is an attempt to generalize, to distill the
factors and processes that drive the behaviors, population
dynamics, and community patterns that we observe. The
strengths of this approach are that the results apply generally to many systems, and that the model allows us to
identify the workings of the hypothesized drivers. Mathematical modeling forces us to be explicit about our assumptions and about the ways that we envision the factors
to be related. Since we often make these assumptions anyway, writing a model almost always focuses our thinking.
We all use models to organize our observations, although
these are usually verbal generalizations rather than mathematical equations. The act of writing an explicit model
forces us to be more precise about the logical progression
that produces particular outcomes. Models also allow us
to explore the bounds of the hypothesis. In other words,
under what conditions does the hypothesis break down?
For example, Bertness and Callaway (1994) developed a
graphical model that predicted that facilitation would be
more frequent when consumer pressure or abiotic stress
was strong and competition would be more frequent when
these stresses were weaker.

Models can be general or specific; both kinds are usually
constructed of mathematical statements. General models

describe the logical links between variables, though they may not contain realistic parameter values. Models that are specific to particular systems often involve measured parameters from actual organisms and allow us to make precise predictions (e.g., how much harvesting can a population sustain?). For example, Kataro Konno (2016) parameterized a simple tritrophic model (plants, herbivores, predators) and found that he could predict the approximate densities of herbivores and carnivores for several terrestrial systems.

Successful models can let us develop new hypotheses about how nature works or about how to manage ecological systems. For example, as mentioned earlier, observational studies had already suggested that competition for resources could affect the abundances of two species at the same trophic level. Theoretical models revealed that the presence of a shared predator (or parasite) could also cause a similar outcome (Holt 1977; see example, figure 1). Partly as a result of these models, Holt and others have looked for this phenomenon, dubbed apparent competition, in nature, and it is indeed widespread (reviewed by Holt and Bonsall 2017).

Models have also proven useful in designing conservation and management strategies. A detailed demographic model of declining loggerhead turtles indicated that populations were less sensitive to changes in mortality of eggs and hatchlings and more sensitive to changes in mortality of older individuals than conservation ecologists had realized previously (Crouse et al. 1987). This result prompted changes in the efforts to protect turtle populations, which has improved their prospects for survival (Finkbeiner et al. 2011, Valdiva et al. 2019).

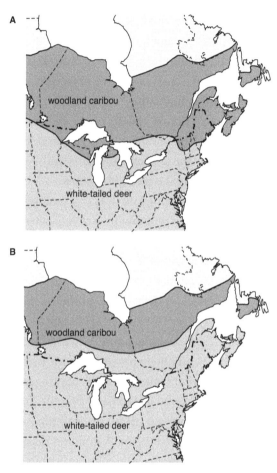

FIGURE 1. Apparent competition between white-tailed deer and caribou (Bergerud and Mercer 1989). A. Caribou historically lived in New England, Atlantic Canada, and the northern Great Lakes states (redrawn from a Wildlands League caribou range map). B. Since European colonization, deer have expanded their range into these areas and replaced caribou (redrawn from Thomas and

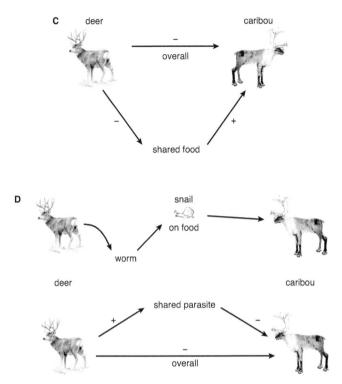

Gray 2002). Numerous efforts to reestablish caribou into areas where they could contact deer have failed. C. The conventional hypothesis for the declines in caribou involved competition for food resources. More deer meant less food for caribou (shown as a negative effect of deer on shared food). This explanation has not been supported by data. D. One alternative hypothesis involves mortality to caribou caused by a shared parasite, a meningeal worm (Vors and Boyce 2009). White-tailed deer are the usual host for the worm, and they are far more tolerant of infection than are moose, mule deer, and especially caribou. Caribou get the worms by ingesting snails and other gastropods that adhere to their food. The gastropods are an intermediate host for the worms.

In addition to helping us develop new hypotheses, models can specify how to test existing ones. Charles Darwin missed an opportunity to test the hypothesis that finches on the Galápagos Islands had differentiated and speciated. On his visit, he observed many finches with different beak shapes and sizes, but he didn't record which morphologies were found on which islands. He didn't see this information as relevant at the time because he had not yet generated the hypothesis (model) that morphologies were adapted to particular environments and that these different species could have evolved on the islands from a single colonizer.

Models can make logical connections easier to see. Often patterns are well known and very visible, but the processes that caused those patterns are difficult to assess. Models can also force us to consider alternative mechanisms when the currently favored explanation does not produce the "right result" in our modeling effort. John Maron and Susan Harrison (1997) wanted to explain why high densities of tussock moth caterpillars were tightly aggregated. They knew from a caging experiment that the caterpillars could survive outside of aggregations. Spatial models suggested that very patchy distributions could arise within homogeneous habitat if predation was strong and the dispersal distance of adult moths was limited. As a result of these model predictions, they looked for this counterintuitive process and found that it was indeed operating.

Models of all sorts have been used to test ecological hypotheses. Models of particular ecological mechanisms can be used to ask whether the model fits actual data. Be careful, though: just because the fit is good does not guarantee that the ecological mechanisms included in the model are operating. A good fit just indicates correlation. Other processes could also produce good fits to the data and may be the actual causal factors in nature. A manipulative experiment would likely be needed to determine causation. Models can be very useful as long as you understand the assumptions that underly them and you don't overinterpret them.

Many of the most influential ecological models are very simple. They ask, "What is the minimum level of complexity or magic that is needed for the world to look the way it does?" When data do not fit these simple models, we get a better understanding of the magic required. The greatest value of models may be not their ability to explain the world, but rather their failure to do so. George E. Box famously said, "All models are wrong, but some are useful."

It is unfortunate that modeling and natural history often attract different people, with different skills and little appreciation for each other's approaches. However, some of the most successful ecologists have been able to bridge these approaches. You don't have to be a master of both— instead, you may get a lot out of teaming up with collaborators who have similar interests but approaches that are complementary to yours.

Why ecologists like experiments so much (or why we couldn't call this book *The Tao of Ecology*)

The Tao is a concept originating in China that refers to the stream-like flow of nature. Like a stream, the Tao moves gently, seeking the path of least resistance and finding its way around, without disturbing or destroying. A Tao of Ecology might entail noninvasive and nondestructive observations of entire systems to understand who the players are and how they interact with one another. We find the Tao to be an appealing concept in general and one that could be applied to ecology. However, nothing could be farther from the approach that many ecologists favor. In this section, we explain why ecologists like to manipulate their systems so much.

When conducting a manipulative experiment, the investigator disturbs the system and observes the effect. This experimental approach has the advantage of providing more reliable information about cause and effect than do more passive methods of study. Understanding cause and effect is critical, powerful, and much more difficult than it sounds.

Consider the inferences that can be drawn from observations versus those from experiments. Observations allow us to make logical connections based on correlations. However, correlations provide limited insight into cause-and-effect relationships. One version of an old adage says that correlation does not imply causation. Bill Shipley (2000) points out that this is incorrect. Correlation almost always *implies* causation, but by itself, cannot *resolve* which of the two (or more) correlated variables might have *caused* the other. Let us give two examples, the first from one of our life experiences and the other from the ecological literature.

The end of grad school was a time of reckoning for Rick. The only car he had ever owned, a Chevy Vega, was clearly falling apart, although he pretended not to notice. His girlfriend at the time convinced him that since he had a job lined up on the other side of the country, and he would soon actually have a salary, he should abandon his grad-student lifestyle and buy another car before heading west. Red has always been his favorite color, so naturally he was interested in a red car. However, his girlfriend had seen a figure on the front page of *USA Today* indicating that red cars are involved in more accidents per mile than cars of other colors. Concerned about their safety, she argued for another color: After all, statistics don't lie, and red cars are more dangerous than other cars. Her working hypothesis had the cause and effect as "red causes danger":

$$\text{red} \longrightarrow \text{danger}$$

Rick was unsuccessful in convincing her that more dangerous (sexy?) people chose red cars in the first place and that getting a more boring color would do little to help them:

$$\text{danger} \longrightarrow \text{red}$$

In the end, Rick bought a grey car, but drives a red one now (when a bicycle won't do). As the third edition goes to press, he has luckily escaped being in any automobile accidents.

This scenario may seem silly and scientists may be unlikely to make this argument (Rick's girlfriend was a social worker). We can assure you that we have seen this type of reasoning repeated many times by ecologists who infer causal

links from correlations. For example, Tom White made the
insightful observations that outbreaks of herbivorous psyllid
insects were associated with physiological stress to their host
plants and these outbreaks followed unusually wet winters
plus unusually dry summers (White 1969):

$$\text{unusual weather} \longrightarrow \text{plant physiological stress} \longrightarrow \text{psyllid outbreaks}$$

He argued that plant physiological stress increased the
availability of limiting nitrogen to the psyllids he studied,
and to many other herbivores (White 1984, 2008). So, essentially, he hypothesized a causal connection between these
correlated factors:

$$\text{weather} \longrightarrow \text{plant stress} \longrightarrow \text{increased N} \longrightarrow \text{herbivore outbreaks}$$

However, the actual causal links could be different. For
instance:

$$\text{weather} \longrightarrow \text{herbivore outbreaks} \longrightarrow \text{plant stress} \longrightarrow \text{increased N}$$

Or perhaps weather influences some other factor that then
causes herbivore outbreaks, without involving the host plant:

$$\text{weather} \longrightarrow \text{plant stress}$$
$$\searrow \text{reduced predation} \longrightarrow \text{herbivore outbreaks}$$

Without manipulative experiments, it is difficult to establish which of these causal hypotheses are valid and important. However, if microenvironmental conditions, physiological stress, available nitrogen, herbivore numbers, and predator numbers can all be manipulated, it will be possible to determine which of these factors cause changes in which others. It is easy to be critical of White's reasoning, although his intuition got him fairly close to what current evidence suggests. A review of experimental studies suggests that herbivores, especially the sap-feeders that White studied, are negatively affected by continuous drought stress but that intermittent bouts of plant stress and recovery promote herbivore populations (Huberty and Denno 2004).

Unfortunately, many ecological problems are not amenable to manipulation. Techniques called structural equation modeling have been developed to provide inference about causal relationships from observational data and will be discussed more fully in chapter 5. These techniques involve directed graphs (the diagrams with arrows shown above). Once we have specified a causal path or directed graph, we can predict which pairs of variables will be correlated and which pairs will be independent of one another. These techniques allow us to build models that estimate the probability of causation from correlations in the data. We can then discard causal models that don't fit our observations.

Structural equation modeling is not difficult to use, although it is less well known than inferential statistical methods such as analysis of variance (see chapter 4). The correlational approach is most useful when one model matches the observed patterns more accurately than alternative models do, which is often not the case in ecology.

In summary, ecologists love manipulative experiments because we love understanding causality. Regardless of your philosophical persuasion about this issue, the truth remains that it is easier to publish experimental work than studies relying solely on observations and correlations. However, not all experiments are created equal. Experiments are only as good as the decisions that stimulated the experimenter to manipulate the few factors that they have chosen. Experiments are also limited by problems of scale and realism.

So, manipulative experiments are powerful, but they can't do everything. Since quantitative observations can be done over large areas and long time frames, they can give a more heuristic perspective than an experimenter can achieve with manipulations. Modeling can provide generality, suggest results when experiments are impossible, project into the future, and stimulate testable predictions.

You are after the best cohesive story you can put together. Whenever possible, you should integrate several of the approaches that we have discussed to pose and answer ecological questions, because one approach can make up for the weaknesses of another. The best research projects tend to combine observations, models, and manipulative experiments to arrive at more complete explanations than any single approach could provide.

CHAPTER 3

Testing Ecological Hypotheses

This chapter describes how to go from having a question to setting up a manipulative or observational study. Chapters 4 and 5 will go into more detail about these two study approaches.

The first step in doing ecological research is to have a clear question or hypothesis in your mind.* If you just have

---

* A historical footnote. The information in this paragraph also appeared in the first edition of *How to Do Ecology*. It became part of official testimony given by Marc Mangel in a case heard by the UN International Court of Justice in The Hague. Japanese whalers have hunted whales for their meat and oil since at least the sixteenth century. However, this practice was made illegal by the International Whaling Commission in the 1980s when the CITES treaty named many whale species as endangered. The Japanese government got around these statutes by arguing that their whaling constituted scientific research, which was exempt from the ban. Since much of this modern whaling took place in the southern Pacific and Antarctic Oceans close to Australia, the Australian government sued at the International Court of Justice to stop the whaling. They argued that the research exemption did not apply in this case, since collecting data was not sufficient to advance ecological knowledge without addressing a specific scientific question. The International Court of Justice ordered whaling "research" to cease. Japan dropped out of the treaty, and continues to whale, though no longer under the guise of ecological research.

a vague interest in a system, an organism, or an interaction, it's a great time to poke around so that questions start to form in your mind (see chapter 1), but you are not ready to start collecting data in earnest. Without a clear question (even if it's a preliminary one), there is no end to the data (relevant or otherwise) that you may feel compelled to collect. For these reasons, you should walk the fine line between pursuing your clear, focused question and being willing to rephrase your question in response to new observations.

A clear question demands that you figure out an appropriate response variable—that is, what you will actually measure to answer the question. Before you start, decide what you are going to measure and how you are going to take your measurements. This step can be challenging for new students. When Rick was starting his thesis, he knew he was fascinated by the extreme life history traits of periodical cicadas and the effects of crowding and predator satiation on their ecology and evolution. He had only a vague idea of his question within this broad area and had no clearly defined idea of what he would ask or measure. After several meetings with his thesis committee, he settled on asking about the effects of crowding on cicada fitness. This allowed him to decide on his response variable, the number of offspring per individual adult (a proxy for fitness) produced at sites with different densities. Remember that after you start working and gain more insight into your system, you may want to reevaluate your question and your response variables.

A clear question stimulates relevant experimental manipulations (or quantitative observations) rather than the other way around. Imagine that you observe pandas and bamboo snout weevils sharing habitat. It does not make sense to begin manipulating the system just to see what happens. Instead, begin by developing your question and only then ask yourself how to test it. So suppose you develop a testable working hypothesis: The density of weevils will be lower in the presence of pandas than in the absence of pandas. Because you have a clear hypothesis in mind, you are now ready to design an experiment. You can conduct a manipulation, excluding pandas from half of your plots and keeping the other half as unmanipulated controls. You can then measure densities of weevils in these two treatments and compare the difference that you measure against the difference that would be expected by chance. Based upon a statistical analysis, you can determine the likelihood that your null hypothesis ("pandas don't reduce weevil densities") should be rejected. Developing a clear question first made all that followed relatively straightforward.

Study requisites

Once you have chosen which response variables you want to measure and what might affect them, you are ready to design your study. A well-designed study must satisfy four requisites, and these apply to both manipulative experiments and observational tests of hypotheses. There are few issues on which all ecologists seem to agree and fewer still that they maintain with religious fervor. Strict observance of study requisites is one of those rare examples. Unambiguous interpretation

of causation is dependent on these requirements: (1) meaningful treatments, (2) replication, (3) appropriate controls,
and (4) independence, randomization, and interspersion
of treatments (Hurlbert 1984).

Meaningful treatments

Establishing cause and effect by a simple manipulative
experiment sounds straightforward but can be surprisingly
difficult. For example, if we want to evaluate the effects of
herbivores on plant traits, we could set up the following replicated and randomly assigned experimental treatments:
(1) plants caged to exclude herbivores and (2) control
plants that lack cages and can be accessed by herbivores.
This experiment seems meaningful and easy to interpret.
However, when we impose experimental treatments, we
often change things other than the factors we are hoping to
manipulate. In this example, any differences that we observe between these two treatments could be caused by the
presence or absence of herbivores, or could be caused by the
cages themselves (a confounding factor). The cages may
alter the microenvironment experienced by the plants; exclude beneficial organisms (pollinators); eliminate harmful
organisms (plant pathogens or other herbivores); interfere
with the normal behavior of the herbivores so that their
effect is greater or less than it would be without a cage; or
interfere with the normal behavior of the plants so that
their usual developmental or reproductive schedules are
disrupted. The list goes on seemingly forever.

There are several ways to minimize the artifacts of
the treatment in this example. You should design a cage

treatment that causes as few of the unwanted, secondary effects as possible. Cages could be constructed of different mesh sizes that allow some of the smaller organisms to come and go freely but block larger ones. Alternatively, cages can sometimes be left open at the top or bottom so that some of the microenvironmental side effects caused by the cage are included in the controls.

It is often a good idea to attempt to impose the treatment in several different ways. Each of the different impositions may have its own side effects. So, for instance, another way to exclude small herbivores is to treat plants with selective pesticides. Such a treatment is likely to cause its own artifacts. However, the artifacts associated with pesticides are probably different than those caused by caging. If you find that herbivores have a consistent effect on plants regardless of how they are experimentally manipulated, you can feel more confident that your conclusions are real and robust.

In addition to diversifying, it's a good idea to try to include the potential side effects of your treatment in controls. If you apply a pesticide using a surfactant and water, spray your controls with a surfactant and water as well. If you inoculate birds against a disease, use a placebo with your control birds so that they have had a similar experience.

Care should be taken in deciding on appropriate treatments and controls. Biological processes may or may not be easily mimicked by manipulative treatments. Consider the example of an experimental treatment that attempts to mimic the effects of fire on plants. Some ecologists mimic landscape-scale fires with small-scale experiments by conducting fires in 1-m² fireproof arenas placed around vegetation. These experimental fires only get a fraction as hot as real fires, combust a small proportion of aboveground biomass, are generally conducted in a different season than wildfires, and so on. If you can't do a larger-scale fire treatment yourself, you may be able to find a collaborator; this will likely require more initiative but may make your study more realistic. Similarly, the easiest and most straight-forward way to mimic herbivores that chew foliage is to cut leaves with scissors. For some plants, clipping with scissors adequately captures the effects of herbivores. For example, damage inflicted by a robotic worm caused many of the same responses in lima bean leaves as actual caterpillars (Bricchi et al. 2010). However, how the leaf area is removed, in one big bite or many small ones, whether the veins are severed, and so on can greatly influence the effect of clipping on the responses of plants (Baldwin 1988). In addition, spe-cific herbivore behaviors and components of saliva from certain herbivores have been found to have profound ef-fects on their host plants (Acevedo et al. 2015). Carefully designed treatments involving actual organisms and actual processes, when feasible, are generally preferable to more artificial treatments.

When designing experimental treatments, it is often important to attempt to span the natural range of varia-tion. Let's say we want to evaluate the disturbances on intertidal invertebrates caused by storms and large drift-wood. We might be tempted to scrape all of the sessile barnacles and mussels from the rocks. However, this ma-nipulation will tell us little about the real effects of patchy disturbance on the intertidal. Conversely, picking only a single, modest level of damage may underestimate the effects of real storms. The best bet here might be to use treatment levels that span the range of naturally occurring damage. Estimates of the natural range may be available in the literature, from your preliminary results, or by asking people familiar with your system, although you may just need to guess. However, some questions may require you to include treatments outside of the current range, for example, for hypotheses about past or future conditions.

Unlike manipulative experiments, observational studies do not allow the experimenter to impose treatments; however, establishing meaningful treatment and control groups or gradients for comparison is essential for them as well.

Replication

It is important to replicate independent units of each treatment and control so that you can separate the effects of the treatments from background noise. Imagine for a moment that there is only one replicate (independent sample) of a treatment. It will be impossible to determine whether any differences between the treatment and a control are really due to differences caused by the experimental manipulation or rather to other differences between the treatment and control. With only one replicate, no amount of subsampling or measurement precision will help establish causality because the factors that affect one subsample may also affect others. However, if many independent replicates of each treatment and control show a difference, we can be more confident that this effect was caused by the treatment.

The same need for independent replicates applies to observational tests of hypotheses. Replication allows you to have more confidence in the causal relationship that you hypothesize is operating in observational studies. It allows you to ask whether factors correlated to your explanatory factors could be causing your results. For example, Dave Reznick and John Endler (1982) observed that guppies from a Trinidadian stream with high risk of predation from larger fish had different life histories than guppies from a stream with low predation risk. Guppies with high risk of predation became adults more quickly and devoted more resources to reproduction than those with low risk of predation. Collecting many guppies (subsamples) from one site of each treatment (high and low predators) did little to improve the inference that predation was the cause of the life history differences. Instead Reznick increased replication in both an observational survey and a manipulative experiment. First, he found many different streams in Trinidad and characterized the life history traits of the guppies as well as their risk of predation. Replication of this observation gave him more confidence that the relationship between predation and guppy life histories was a real one. Second, he moved guppies experimentally from streams with high predation to streams with low predation (Reznick et al. 1990). Descendants of the transplanted guppies had traits that matched those of the guppies that had lived in low predation streams for generations. Furthermore, this study was repeated in streams in two different river systems. This work is convincing because the results were consistent across many spatial replicates.

Often replication comes at the expense of precision of each measurement. It may be counterintuitive, but this lack of precision is not a problem. It is almost always better to take many samples, each messy (i.e., with little precision), than to spend your limited time making sure that your measurement of any given sample is exact. The central limit theorem of probability can help you out here. If you take a large number of unbiased estimates, each one very imprecise, you will quickly arrive at an estimated mean value that is close to the actual mean value. This is an impressive party trick. Get your friends to estimate the size of some object, say a window. Individual estimates are likely to be far from the actual value (geez, do some people have bad judgment). However, the mean from a group of about 30 partyers will be astonishingly close to the actual value. The message from this exercise is clear. Always go for as large a sample size as you can get, even if each of your samples is sloppy and noisy. A large, unbiased set of samples will average over the noise and bail you out. This advice is supported by analyses of real ecological data (Zschokke and Ludin 2001). Imprecise measurement had surprisingly little effect on ecological results, suggesting that limited time and resources are better invested in more replicates than in greater precision of measurements. This is a hard idea to digest but will make your data collection much more effective if you take it to heart.

A large sample size can rescue imprecise measurements, but it cannot rescue biased measurements. For example, imagine asking a group of 30 partyers for their estimates of the age of the earth. The mean value will likely be quite

different if the partyers are geologists or Christian fundamen-
talists. Increasing the sample size will not alter this bias.

The key point is that, in general, the more independent
replicates you have, the greater your power to detect treatment effects. Students who are just starting to do research
often want to know how many replicates they will need.
There is no easy answer to this question; it depends on the
size of the difference (effect size) that you are interested in
detecting and how much noise there is. Yes, but let's get
real. That knowledge doesn't help you determine how large
a sample size to shoot for. As a general rule of thumb with
no other information about effect size or noise, we try to get
30 independent replicates of each treatment or category
(discrete variable). If 30 is impossible, you may be able to
get by with 15. Below 15, ecologists start to get anxious.

Some studies do not lend themselves to high replication. For example, conservation questions at the land-
scape scale can usually be replicated only a few times. We
have conducted experiments in large plots that excluded
different mammalian herbivores (Young et al. 1998). In
these cases, each treatment was replicated only three times.
Since we were looking for relatively huge effects, we were
able to detect significant differences even with low replication (Huntzinger et al. 2004).

Sometimes it is impossible to replicate your treatments
at all. In these cases, traditional interpretations of statistical
tests are probably inappropriate (although this suggestion
is contentious—see Oksanen 2001). Large-scale results without statistical tests are difficult to publish by themselves but

may accompany smaller-scale replicated studies to provide biological realism.

So far, we've mentioned only spatial replication. What about replication through time? There's at least as much noise through time as there is across space. Conditions such as weather, fire or flood events, and species outbreaks and crashes vary enormously from season to season. Imagine if our discipline required on the order of 30 temporal replicates and each replicate required one year. Achieving temporal replication is so challenging that ecologists have largely punted on it. This state of affairs is almost certainly the result of logistical convenience rather than because temporal replication isn't important. In reality, funding and our expectations for productivity make it difficult to replicate experiments over more than a few years. Yearly variation was found to have strong effects on the results of ecological experiments; nonetheless, temporal replication is rarely addressed in the design or interpretation of experiments (Vaughn and Young 2010).

BALANCING SCALE AND REPLICATION

Having a large number of replicates increases your power to detect differences caused by your treatments. However, high replication comes at the expense of the scale of each sample. In other words, if you want to have many replicates, each of those replicates is going to be small; if you have fewer replicates, each one can occupy a larger area. This can be a serious problem because some processes only operate at particular (e.g., large) spatial scales.

Insufficient replication is just one possible source of an incorrect interpretation. Experiments conducted using spatial and temporal scales that are too small can also lead to incorrect inferences. Since replication almost always comes at the cost of scale, some ecologists argue that our field has leaned too far in the direction of replication and that scale should take priority (Oksanen 2001). In addition, resource managers, growers, and other stakeholders are often not very hung up with statistical tests (which require high replication) but won't listen to results conducted on small plots. On the other hand, as we've mentioned, most academics will not listen to results (or publish them) if they are not statistically significant, and this requires replication. And you can't achieve both goals simultaneously. The solution? Conduct two different tests, one with high replication and the other with a large spatial scale. It probably doesn't matter which scale you start with; there are unique advantages to each. If the answers are similar at both scales, the conclusion is much stronger.

One way to visualize the scope of your experiments and observations is to plot them on a graph with spatial scale as one axis and temporal scale as the other. In this way you can clearly see the spatial and temporal range that your manipulative experiments, quantitative observations, and models cover. For example, Schneider et al. (1997) wanted to understand the multiannual population dynamics of a bivalve mollusk at the scale of an entire harbor (368 km^2). Their experimental units were 13-cm cores taken over a 30-second period. These units were repeated over an area of about half a kilometer^2 during a 28-month period, which

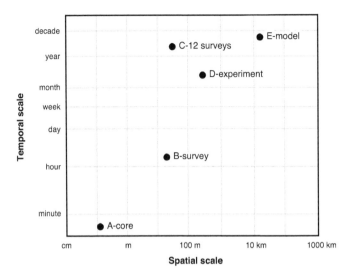

FIGURE 2. A scope (scale) diagram showing observational, experimental, and modeling approaches used to characterize changes that occur to harbors over decades (Schneider et al. 1997). The diagram makes explicit the spatial and temporal extent of each observation or experiment, as well as the scale over which Schneider et al. would like to extrapolate. The observational and experimental unit for all sampling was a 13cm-diameter core taken for 30 seconds (A). A survey included 36 cores at six sites (B), repeated during 12 monitoring visits (C). An experiment was conducted in which 10 cores were taken at each of two experimental plots, each of which was visited nine times over a 100-day period (D). Schneider et al. also constructed a model at the scale of the entire harbor (E) and attempted to extrapolate downward to compare their model results with those from observations (surveys) and experiments.

allowed them to greatly expand the scope of their experiment. Nonetheless, making inferences about dynamics that occur in the harbor on the scale of decades required considerable extrapolation (figure 2). They combined this experiment with a model of the entire harbor. They used information from the small-scale experiment to suggest model parameters and information from the model to help interpret their experimental results and suggest further experiments.

Since manipulative experiments in ecology are almost always conducted at spatial and temporal scales smaller than our ideal, it is worth considering what effect this has on our worldview. Small-scale experiments have led us as a group to believe in local determinism, that is, that the processes we can manipulate on a small scale mold the patterns that occur at larger scales (Ricklefs and Schluter 1993). However, this belief is likely to be simplistic when we look at real communities. For example, local processes such as competition and predation tend to reduce species diversity, whereas larger-scale regional processes tend to increase diversity through movement and speciation. One way to see that small-scale experiments don't capture all of the important processes is to place a barrier around a local area and observe whether all of the species persist. In most cases, they don't. Even the largest parks, such as Yellowstone and the Serengeti, are too small to maintain a full complement of species over the long term (Newmark 1995, 1996).

In summary, if at all possible, replicate your treatments. Doing so will increase your confidence in the cause-and-effect relationships you find (not to mention making your results more publishable). However, remember that replication often comes at the expense of scale. When you interpret your results, be mindful of the actual scale of your study and extrapolate cautiously.

Controls

Controls are organisms or plots against which manipulations and other treatments are compared. Consider a biological system that changes over time. Any differences that we observe before and after application of the treatment could be caused by the treatment but also could be caused by other changes that occurred during this time period. However, if we have controls that don't receive the treatment, we can separate the effects caused by the treatment and those that are unrelated to the treatment. If we wanted to understand why male deer shed their antlers, we could experimentally add French fries to the diets of male deer during fall and winter. We would observe that they shed their antlers in March. Without controls, we might conclude that the added experimental fries had caused the antlers to be shed. However, controls without added fries would help us recognize that other seasonal changes were responsible for the antler shedding that we had observed. In this case, our control treatment consists of male deer with no added food, essentially "no treatment."

The logical category or control in some experiments is not necessarily "no treatment." For example, if we wanted to evaluate effects of elevated CO_2 on plant growth, we might want to compare plants grown in an environment with CO_2 levels that are projected for 2050 with controls grown under current ambient conditions (no treatment). However, more meaningful controls might be plants grown under conditions with 25% less CO_2 than current ambient conditions, since this is the estimated level before the Industrial Revolution.

Independent replicates, randomization,
and interspersion of treatments

Replication serves a useful purpose in study designs only if the replicates are spaced correctly. For instance, if all of your high-nitrogen replicates also happen to be in a swampy area and your controls are in a drier upland area, then you might conclude that high nitrogen caused effects that were actually caused by the wetland. Independent replicates make it likely that the different treatments are similar in all ways except for the treatment effect.

Consider a design for an experiment to examine the effects of ladybugs (biological control agents) on greenhouse insect pests. The greenhouse is physically divided in half with a screen barrier down the middle (figure 3A). Predators of the pest are released into one half of the greenhouse, containing, let's say, nine plants. No predators are released into the control half of the greenhouse, also containing nine plants. It is incorrect to assume that each of the two treatments is replicated nine times. If one side of the greenhouse is different from the other, then all of the plants of each treatment will experience those differences. In essence, there is only one independent replicate of both the treatment and the control, each of which has been subsampled nine times.

The way to create independence of replicates is to intersperse the replicates of the various treatments. In other words, placement of the treatments and controls must be all mixed up (figure 3B). If one side of the greenhouse is sunnier or windier than the other, these differences will not be confounded with treatment effects. Any observed

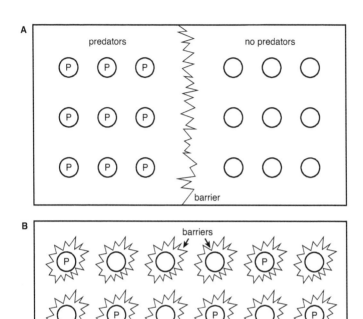

FIGURE 3. Experimental designs to evaluate the effects of predators on greenhouse pests. A. A pseudo-replicated design with one barrier separating the two treatments and only one independent replicate of each treatment. Treatments are not interspersed. B. A proper design with 18 barriers and nine independent replicates of each treatment.

differences associated with the treatments will probably be caused by the treatments. It is unfortunate, but this design is going to require many more screen barriers than the design that divides the greenhouse in half. In summary, by having treatments interspersed with controls, we can differentiate

between the possibility that our treatments caused the results that we observed and the possibility that some other
factor caused the results.

Getting independence between replicates is not always
easy. In practice, try to separate replicates by enough space
so that conditions for one replicate do not influence relevant
conditions for another replicate. This distance will be determined by the organisms—generally, larger or more mobile organisms will need greater spacing than smaller, more
sedentary ones.

The best way to ensure that treatments are interspersed
is by randomizing them. Random assignment does not mean
assigning every other individual to each treatment the way
you deal cards, nor does it mean going along and haphazardly assigning treatments. Both of these methods are ac-
ceptable, but they should be identified as an alternating
(regular) or a haphazard assignment of treatments, and both
will reduce the power of your statistical inference. In summary, randomization is your most powerful option.

Randomization is usually an effective way to achieve interspersion of treatments. To assign your treatments randomly, use a free, online random number generator. Make
sure that the same number of replicates are assigned to
each treatment. Sometimes the process takes a long time,
so it's good to take care of this before heading into the
field.

Randomization is useful but not foolproof: it can fail to
provide treatments that are well interspersed and matched
for other factors. If this occurs, the experimental units
should probably be reassigned (Hurlbert 1984). In other

words, if, by chance, many of one treatment wind up being on one side of the plot (which turns out to be drier) and many of the second treatment wind up being on the other side of the plot (which turns out the be wetter), moisture and treatment will be confounded. Such effects may not be obvious, so it is worth assigning treatments randomly a second time; even if you don't suspect that the two sides differ, there are many confounding spatial factors that you might not anticipate.

SPATIAL NON-INDEPENDENCE

If you know that there is an environmental gradient in your study plot before you assign treatments, it is a good idea to block (split) your plots and assign treatments within blocks (Potvin 1993). For example, if you know that one part of your plot is wet and the other part is dry, divide the plot into two blocks (wet and dry) and then randomly assign an equal number of replicates of each treatment to each of these two blocks. Blocking can reduce the noise caused by the environment and give you more power to detect an effect of your treatments when the blocks are very different from one another. However, there is a cost of blocking. Blocks decrease the statistical power of your analysis—the more blocks, the larger this decrease in statistical power. Blocking only adds power if the blocks accurately match the environmental heterogeneity that is important to the outcome of your experiment (e.g., wet side vs dry). If the blocked factor is not important, then blocking decreases power relative to a completely randomized design.

Two samples that are close together in space are less likely to be independent because the same single event may have affected both of them, an issue called spatial autocorrelation. If you know where samples are located, it is possible to test for similarity based on distance and to remove this effect using a spatially explicit generalized least squares analysis (Legendre et al. 2002, Dray et al. 2012). This is a form of multiple regression that tests whether samples that are close together might be similar for reasons that have nothing to do with the treatment. Perhaps you wish to determine the predation risk of individual coral polyps subjected to heat treatments. If polyps that have the lowest survival rates are next to each other, it is possible that the same predator grazed on both of them out of convenience, independent of the heat treatment. A spatially explicit model will weight these observations less than if the coral polyps with the lowest predation rates are found spread across the survey area.

Mixed or multilevel models can allow you to estimate how important non-independence may be. These models evaluate sites as a random effect, allowing you to compare the importance of spatial non-independence relative to the importance of your treatments (fixed effects) on your response variable. In other words, they suggest how much of a problem the lack of true replication is in your design.

Let's assume that instead of one greenhouse in the example given earlier (figure 3), there are 20 greenhouses, each with 10 plants. Half of the greenhouses (10 greenhouses, 100 plants) could be assigned to have ladybug predators and half (10 greenhouses, 100 plants) no predators. Since the plants in each greenhouse are not independent of each other, the number of independent sampling units would be 20 and not 200. All of the plants in a single greenhouse might be affected in the same way by whatever weirdness goes on in that greenhouse—a difference in water, temperature, diseases, other insects, and so on. If you include the identity of the greenhouses as a random effect in your multilevel model, you can estimate how much of the variance in the number of insect pests on each plant is due to the ladybug treatments (the fixed effect you care about) and how much is due to differences among greenhouses (the random effect you don't care about).

This approach only works well if you have a relatively large number of categories of the random effect—in this case, 10 greenhouses of each treatment. Statisticians differ in their recommendations for how much replication is required for the random effect (how many greenhouses or categories), although the recommendations tend to be in the range of 5–15 as a minimum. In addition, applying mixed models correctly can be complex and has the potential to be abused (Arnqvist 2020).

PHYLOGENETIC NON-INDEPENDENCE

Spatial non-independence is just one of the situations that can bias results. A shared evolutionary history can also make replicates non-independent. For example, perhaps you want to compare the feeding habits of lizards in desert scrub surrounding the Gulf of California with lizards in similar habitats surrounding the Mediterranean. You might observe that many North American lizards have prehensile tongues and often eat insects. Lizards from the Mediterranean lack these traits. You might conclude that something about the Americas selected for these traits repeatedly. However, many of the American species are related iguanian lizards and these traits are generally restricted to this group. The lizards from the Mediterranean are not iguanids. So, much as in the greenhouse situation above, a sample of the dozen or so lizard species from each location does not compare phylogenetically independent species from these two environments (Vitt and Pianka 2005). Any differences in the feeding habits of lizards from these two locations probably results from their shared evolutionary histories rather than from the particular ecological conditions where they are currently found. But it is sometimes possible to make comparisons among species that take shared phylogeny into account. For example, when Anurag Agrawal and Peter Kotanen (2003) designed a study to compare how much herbivory native versus non-native plant species experienced, they chose pairs of species within the same genus or family where one was native and the other was introduced. In this design, each taxonomic group was equally represented in both treatments.

Shared phylogeny can also inform questions about the origin and adaptive significance of traits. Comparisons among taxa can allow you to identify broad-scale patterns spanning long periods of time (Weber and Agrawal 2012). For instance, plants from very diverse lineages (families) in arid environments have repeatedly evolved succulent leaves and stems. Manipulative experiments will rarely be able to address questions covering these large taxonomic and temporal scales.

You can account for the non-independence of shared ancestry using phylogenetically explicit generalized least squares analyses and several related techniques (Garland et al. 2005). These analyses require a well-supported phylogeny (history) to remove the structure of evolutionary relationships from the model. Ian used this method in his analysis of relationships between caterpillar performance and the leaf traits of 27 oak species (Pearse 2011). Phylogenetic relationships among the oak species had been determined in a previous study (Pearse and Hipp 2009), and a phylogenetically explicit analysis was used to factor out the effects of shared ancestry. Fortunately, most of the plant traits that caterpillars cared about (e.g., tough leaves) had evolved multiple times. If there had been few independent origins of those traits, the study would have had little ability to differentiate effects due to leaf traits from effects due to shared ancestry.

Even the fanciest statistical tools cannot calculate away complete non-independence of samples, and more independent samples will yield results that are more likely to provide meaningful inferences. On the other hand, finding that a pattern has a strong signature due to a spatial or phylogenetic constraint may be interesting in its own right. In any case, it is important to think about the independence of your sampling units and related consequences and to design your study accordingly.

Lab, greenhouse, or field?

Ecologists love to understand causality. What better way to establish simple causal relationships than to work in environments that minimize unwanted variance or "noise"? The reasoning is that controlled environments enable us to vary single factors to isolate their effects. The real world (field) can be so complicated that it can be difficult or impossible to perceive patterns because of all the noise. We minimize noise in different ways. We choose only field sites that are well matched and as similar to one another as possible. We move into the greenhouse, where abiotic conditions can be controlled and made similar for all replicates. Sometimes we conduct experiments in small growth chambers, aquariums, or lab "microcosms" that provide even more environmental control. A simplified, controlled environment can reduce this noise and allow us to see the signal, test predictions, or get at mechanisms that would not be possible in the field. In addition, working in these controlled environments is often more convenient than working in the field; they may be close to where we have equipment or other obligations (e.g., our classes or families) and allow us to conduct experiments during times when natural systems may be inactive.

This control and convenience come at large, and often unrecognized, costs. First, controlled environments are generally far more variable than we imagine (Potvin 1993). In our experience, plants and insects grow very differently on one side of a greenhouse bench than the other. This variance in the greenhouse is often larger than we have encountered in the field. In general, organisms in the greenhouse or in aquariums routinely experience weird conditions, such as pest outbreaks, that do not reflect common conditions in the field.

Second, we are often focused on differences in group means; variation in real organisms around those means can be interesting even if it interferes with our desire to isolate single causal drivers. For example, repeatable behavioral differences among individuals can shed considerable light on many ecological processes (Sih et al. 2012). Ian has considered the spatial and temporal variation found within individual plants in their defensive chemicals rather than measuring only the mean level for each individual. He found that those individuals that were more variable were better protected, i.e., that variability itself provided defense (Pearse et al. 2018).

Third, it is often interesting and informative to compare our treatment effect relative to other ecological effects. If we go to some length to minimize those other effects (noise), we lose the ability to put our experimental effect into a meaningful context. (See "Yes/no questions versus relative importance" in chapter 4.)

Fourth, working under controlled conditions is unrealistic and less likely to provide useful inferences about nature. While Rick was asking questions in the field about induced resistance to herbivores (that is, defenses against chewing that don't ramp up until chewing begins) in wild tobacco plants, an unexpected frost damaged many of these plants. This seemed like a disaster at first, but he learned that induced plants are more susceptible to frost, and this risk may represent an unappreciated cost of induction (Karban and Maron 2001). In contrast, after comparing many repetitions of a lab experiment, Rick found that the strength of induced resistance varied from one experiment to the next. To his disappointment, he figured out that some of this variation was due to using pots of different sizes (Karban 1987). Plants in pots dry down at different rates, and pot-bound plants are less inducible. Who cares? This result provides very little inference about how organisms work in the real world.

Laboratory experiments are usually conducted under conditions that are simplified and controlled by intent. Even if you are able to set up the experiment and answer the question that you posed, you cannot know how well it depicts similar processes in nature. The solution to this problem is to link lab and field studies. They can each provide unique information but also have unique limitations. Field observations and experiments should be followed by lab studies to learn more about the ecological mechanisms that could cause the field result. In turn, lab studies should be followed by field studies and observations to learn whether the lab results are realistic and whether they hold at larger spatial and temporal scales (Diamond 1986).

This chapter has explained how to design an experiment that is tightly tied to your question and that adheres to study requisites. Your design should have meaningful treatments, replication, and controls, and your treatments should be randomized, interspersed, and independent. For manipulative experiments, each of these requisites is required. For quantitative surveys, you may not be able to follow the letter of the law, but you should strive to follow the spirit of the law. That is, choose your plots or focal organisms (your replicates) with these requisites in mind.

CHAPTER 4

Using Statistics in Ecology

This chapter considers why ecologists insist on using statistics to evaluate hypotheses. While you may run across old studies without proper controls, or replication, or even statistics, it has been virtually impossible for many decades to get studies published without statistical analyses. Since human beings are extremely proficient at seeing patterns whether they exist or not, it's important to impartially evaluate the story we think we see. Statistics allow you to evaluate whether the differences caused by your treatments (or groups) are likely to be real patterns or just noise.

Doing and presenting statistics

Statistical vs biological significance

When we observe a difference between two treatments or a relationship between two factors, statistical tests provide an estimate of the likelihood that the difference or relationship was caused randomly. However, we often confuse statistical significance (indicated by a probability level) with ecological significance (indicated by the size of the effect—how different two things are). For example, we might hypothesize that diet affects the body size of rodents. If we feed two groups of individuals different diets, we can be fairly certain that, with enough replication, mean weights of the two groups will not be the same. Very small differences can be statistically significant but not produce consequences that are ecologically important, such as differences in reproductive output.

When we present our results, we must include both statistical significance (represented by p-values; see below) and biological significance (represented by effect sizes; see below and chapter 8). It is insufficient to report that two populations were different and give a probability ($p < 0.05$ or $p = 0.023$, for example) or to report that differences were not significant (ns). Along with any reports of statistical significance (or, in our opinion, near-significance), provide the effect size. This can be done by showing a figure (perhaps a bar graph with the means and standard errors for each of the populations) or by reporting that, for example, grasshoppers were 35% more numerous when chameleons were absent. There is no hard-and-fast rule for interpreting biological significance. A doubling of the number of mountain lions in a particular area may have a large biological effect on the deer that live there, but a doubling of the number of mountain lion carcasses in that same area may make little biological difference to nutrient pools.

p-*values*

Although statistical analyses are essential for ecology to progress as a science, our emphasis on significance tests may be a bit overly zealous (Yoccuz 1991, Berner and Amrhein 2022). By convention, we have decided that we will consider two populations to be different if the probability of them being the same is less than 0.05. But whether we find this magical 0.05 threshold depends on the variation within the population and on our sample size. If we have a very large sample size, two populations can be statistically different with means that are quite close. On the other hand, if our sample size is small, two populations that are actually quite different will not appear significantly different at the 0.05 level. It's baffling that a group of intelligent and thoughtful people can be so hypnotized by this essentially arbitrary number.

We know that no two populations are identical, just as no two people are. When we test a null hypothesis that two populations are the same, we are not calculating the probability that they are truly identical but rather the probability that we can detect a difference between them. That is, statistical significance is really a property not only of the ecological system, but also of the data collected and the experimenter's ability to make distinctions. Unfortunately, the 0.05 threshold has become an absolute wall between "real" results and "negative" results. Why should we be allowed to say that two populations are truly different if $p = 0.049$ but not be allowed to say much of anything if $p = 0.051$? In both cases our inference about the populations being different will be wrong approximately 5 times out of 100 (see below). You should be aware of the arbitrary nature of the 0.05 threshold and interpret results accordingly. Whenever possible, give the calculated p-value rather than reporting that p is greater or less than 0.05. If $p = 0.001$, you can be more confident that the result was not caused by chance than if $p=0.05$. Similarly, your confidence about $p=0.06$ should be different than $p=0.60$.

Biases

How you phrase your question or hypothesis can affect your estimate of statistical significance and the conclusion that you reach. Aschwanden (2015) provides a fun, interactive example asking whether Democrat or Republican governors have been better for the American economy since 1948. The "economy" is a vague and complicated response variable to evaluate. If your response variable was gross domestic product, you would conclude that Democratic governors were better ($p<0.01$), but if you looked at employment numbers, you would conclude that Republican governors were ($p<0.01$). If you looked at the growth of the stock market, you would find no difference ($p=0.27$ to $p=0.33$). The same issue can occur with ecological questions. Researchers wanted to know the possible effects of wolves on elk browsing and aspen growth in Yellowstone. When they considered only the tallest aspen trees as their response variable, they concluded that wolves changed the elks' browsing behavior and ultimately resulted in greater growth of aspens. However, when growth of a random sample of trees was measured as the response variable, the indirect effect of wolves on aspen was much reduced (Brice et al. 2022). What you choose as your specific response variable influences (or, in some cases, outright changes) the conclusion that you draw.

Type I and II errors

Even when we use statistical inference correctly, we sometimes come to the wrong conclusions. For example, Matthews (2000) tested the null hypothesis that (despite the widespread fairytale) there was no relationship between stork numbers and human birth rates. However, he found a strong positive relationship between the number of breeding pairs of storks and human birth rates in 17 European countries ($p=0.008$, figure 4). The cause-and-effect relationship between storks and human birth rates was still unclear. The hypothesis that storks bring human babies gained more support from surveys conducted in Berlin between 1990 and 1999 (Hofer et al. 2004). There was a strong positive, linear relationship between the number of stork pairs recorded in the countryside surrounding Berlin in each year and the number of non-hospital human births in Berlin during that year ($R^2=0.49$, $n=10$, $p=0.024$). Presumably because storks don't bring babies to hospitals, no relationship was found between the number of stork pairs and hospital births ($R^2=0.12$, $n=10$, $p=0.32$). The conclusion that storks bring babies is an example of a type I statistical error, concluding that a relationship exists when, in fact, it doesn't. Ecologists are likely to make type I errors five times out of 100 if we accept a threshold of $p=0.05$. Type II errors occur when we fail to conclude that a factor is significant when, in fact, it actually is. For example, we might conclude that sexual intercourse has no significant relationship to having babies because there are many cases in which intercourse does not result in a baby. Type II errors are probably much more common in ecology (and family planning) than type I errors.

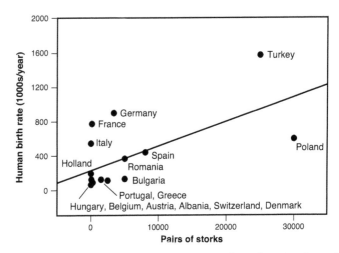

FIGURE 4. The relationship between the number of stork pairs and the human birth rate in 17 European countries, 1980–90 (redrawn from Matthews 2000). Countries with more stork pairs had higher birth rates (births in 1000s = 0.029 stork pairs + 225, r = 0.62, df = 15, p = 0.008).

Alternative hypotheses

Null hypotheses in ecology

Scientists have been urged to design null hypotheses that can be critically tested and rejected (Popper 1959, Platt 1964). This is sound advice, although it doesn't work very well for many ecological questions (Quinn and Dunham 1983). Ecological hypotheses are often not simply true or false. For example, suppose we are interested in asking what the role of competition is in structuring communities. This is not a question that can be falsified by conducting a simple experiment. Evidence for competition can be found in most systems. However, so can other processes like predation and parasitism, facilitation, abiotic factors, and disturbance. In-stead of rejecting hypotheses and asking yes/no questions, we should be weighing alternatives and asking how impor-tant competition is relative to other processes. In other words, we are more interested in discovering the size of the effect caused by competition and comparing it to other drivers than trying to reject a null hypothesis that competi-tion does or does not operate. In ecology, unlike in some other scientific disciplines, the principles are not universal. Finding a single counterexample would make us rethink our working hypothesis about the force of gravity. However, finding a single counterexample does not disprove our ideas about competition. Similarly, ecological hypotheses are rarely mutually exclusive. Even if competition is found to be impor-tant, predation may also be important.

Too often ecologists have become advocates of a polar-ized point of view (e.g., density dependence versus density independence) and then spent their careers tenaciously defending it. Ecologists should strive to explicitly develop alternative hypotheses to explain the patterns they observe (Platt 1964). By developing a set of alternatives, you can avoid becoming emotionally attached to the hypothesis that you selected initially. Rick's son Jesse built a crude lep-rechaun trap in kindergarten, baited it with "gold" rocks, and placed it out on the eve of St. Patrick's Day. The next morning, as Jesse checked the trap, one of the rocks dropped into the grass, an event he failed to observe. He was immensely excited by this "evidence" that the lepre-chaun had visited, taken one of the pieces of gold, and got-ten away. The next year, he built a more sophisticated trap.

Although the leprechaun failed to visit in this second year, he still felt confident of their existence based on the previous season's results. It is unfortunate that Jesse did not consider alternative hypotheses for the rock's disappearance. It's worth wondering how many leprechauns we each catch in our research careers.

Insurance

We are all under pressure to produce significant results. Testing alternative hypotheses is an efficient way to ensure that you have something to say when you get done. If you become enamored with a particular process and conduct experiments focused only on that process and its potential mechanisms, you'll have a story to tell only if that process turns out to be as important as you thought it was. Starting with a list of alternative hypotheses can reduce your stress because you are much more likely to turn up something interesting.

Often after you have tested a hypothetical ecological mechanism, it becomes clear that there are alternative mechanisms that could also have contributed. Our friend Kevin Rice recommends avoiding the "house of cards" research program. Imagine a scenario in which all of the interesting secondary hypotheses require a particular outcome to be true in your initial hypothesis. This puts you under too much pressure to demonstrate your initial hypothesis, whether or not it is actually true. If, instead, you consider alternative hypotheses, you have something to talk about no matter what you find. Box 3 provides suggestions for generating alternative hypotheses in ecology.

Box 3. *Generating alternative hypotheses*

Once you have identified a pattern that is interesting to you, think about a working hypothesis to explain or produce the pattern. Next, consider alternative hypotheses that could also produce that pattern. Try the following list of possible factors as alternatives that could also have produced your pattern:

· abiotic factors (precipitation, temperature, light, fire regime, etc.)
· predators, parasites, and diseases
· competitors and mutualists
· mating factors (sexual selection, nest-site availability, opportunities for offspring, etc.)
· microhabitats (shelters from abiotic conditions, predators, etc.)
· disturbance attributed to human influences or natural causes
· genetic or ontogenetic (developmental) influences
· patterns of dispersal
· historical events

Your list of alternatives can get long and unwieldy, but this is an important step in doing good science. You don't necessarily have to test all of your alternatives, though getting them all down on paper for consideration is a first step. Prioritize them based on how compelling and how testable each one is, and design tests for the ones at the top of the list first. A fun way to generate this list is to use the activity outlined in box 4.

Yes/no questions vs relative importance

Answering yes/no questions will often take the form of rejecting hypotheses, but we have argued that many ecological hypotheses cannot be rejected in this way. Instead, it may be more useful to devise a list of alternative hypotheses, acknowledge that most or all of these working hypotheses may be valid, and then attempt to determine the relative importance of each. This process is akin to partitioning the variance in ANOVA that is due to each of the working hypotheses.

As an example of the process, Rick observed that spittlebugs, plume moth caterpillars, and thrips all fed on seaside daisy along the California coast. He wanted to know how these three common herbivores affected each other. Instead of just testing the hypothesis that they competed (yes/no), he examined the relative importance of interspecific competition, predators, and plant genotype on the success of each herbivore (Karban 1989). This was done by including all three factors (competition, predation, host-plant effects) in one experiment and partitioning the variation in each herbivore's performance (survival, fecundity, population change) that could be attributed to each factor.

Shape of the relationship and ANOVA versus regression

The shape of the relationship (linear or curved) you anticipate between a causal factor and its effects determines the number of "levels" you should include. In the seaside daisy experiment, Rick compared the effects of the complete removal of competitors (or predators) with natural densities of competitors (or predators). In the jargon of statistics, there were only two levels, all or none. This design works best if the effects of the predictor (in this case presence/absence of competitors) on the response variable (e.g., survival) are linear. Unfortunately, ecological effects are often not linear. Examining the natural relationship between numbers of competitors and performance over space or time can provide valuable intuition. When you suspect that the relationship between the predictor variable that you are manipulating and the response variable may be nonlinear, you can adopt a design that involves many levels of the predictor variable (Cottingham et al. 2005). For example, you could assign many different levels of competitors that spanned the entire range of values you observed in nature. This design would be analyzed with a regression rather than an analysis of variance (ANOVA). Regression is similar to ANOVA except that it has many levels (rather than two or a small number in ANOVAs) and it does not require replication at each level, as ANOVAs do. Most regression designs involve linear relationships between variables, although more sophisticated methods (e.g., cubic splines, general additive models) allow you to determine the shape of the relationship in more detail.

Bayesian statistics

In recent decades, ecologists have become interested in Bayesian statistics, in part because they allow us to evaluate how well multiple working hypotheses fit data (Hilborn and Mangel 1997, Gotelli and Ellison 2004). Instead of rejecting a null hypothesis, the result of a Bayesian analysis is an index of confidence in each of several hypotheses.

A Bayesian approach lends itself beautifully to evaluating alternatives and can be a valuable tool for ecologists capable of using it. Bayesian analyses consider all of the possible outcomes, but estimate how probable various explanations are for the observed data given the conditions or information that you have specified. The analysis ranks the various outcomes depending upon their relative likelihoods. Unfortunately, Bayesian analyses require a lot of computational sophistication, and "ecological detectives" without a lot of background and confidence in mathematics may find them daunting. Richard McElreath (2019) offers a course, a textbook, and R code that provides an accessible introduction to Bayesian analysis for biologists.

No matter what statistical techniques you use, experiments that allow you to test multiple hypotheses will often be more effective in ecology than those that test a single null hypothesis.

Negative results and meta-analysis

Earlier we discussed the statistical procedure that lets you reject null hypotheses with varying levels of confidence. When we encounter a negative result or fail to get a significant effect (i.e., when $p > 0.05$), does that mean that the null hypothesis is true? In other words, if we compare two populations and cannot conclude that they are different at the $p \leq 0.05$ level, then should we conclude that they really are the same? The answer to both questions is no. Based on the information we have, the only thing we can conclude is that we failed to find the difference or effect we had hypothesized. Our statistical tests give us far more power to reject hypotheses than to accept negative results as reality. In most cases we have very weak power to evaluate whether two populations are similar. Furthermore, we rarely use statistics to address the question of whether characteristics of two populations are really the same. Many negative results (results that are not statistically significant) never get published, a loss to the scientist who did the work and to the ecological community that never gets to hear about them.

A technique called power analysis is available that allows you to evaluate whether the effect caused by one factor is as great as the effect caused by another factor. This technique allows negative results to become as informative as "positive" results. Unfortunately, larger sample sizes are required to reliably infer that a treatment did not affect a population than to conclude that it did have an effect (Cohen 1988). Our ability to accept a negative result depends on the effect size (the degree to which the treatment means were different). By convention, some statisticians tend to define a small effect size as a difference of 0.10 (10%) or less and a large effect size as 0.40 (40%) or greater (Cohen 1988). We can be more confident that a large effect did not occur than that a small effect did not occur. Cohen (1988) provides a very readable discussion with worked examples of how to calculate the probability that your result of "no significant difference" reflects the actual biological situation in your study. Many statistical packages calculate a value for "power," which tells you how likely you were to find a significant difference given your amount of replication and your observed effect size (whether small, large, or in between). Low power means you should have low confidence in negative results.

A useful way of comparing effect sizes from different studies is meta-analysis (Koricheva et al. 2013). Each study becomes one independent measure of the effect of a particular factor or response variable. For example, we might be interested in the effect of removing top predators on lower trophic levels in a variety of published studies. A meta-analysis lets us formally and statistically compare the results of many studies and can also allow us to put our own experimental results in a much broader context. For each published study, we can calculate an estimate of the effect size by comparing the means of the treatment groups (with and without top predators), standardized by the variance. The meta-analysis can help evaluate whether conclusions from one study are general. It can also provide information about the conditions under which extrapolation from our experimental results is warranted. For example, a meta-analysis revealed that studies of trophic cascades (removal of top predators and ensuing effects) produced larger effects in aquatic systems than in terrestrial systems (Shurin et al. 2002). This general conclusion would not have been possible based on the results of any single experiment.

Meta-analysis can only be meaningful if it includes all available information on a particular question; for this reason, publishing negative results can be critically important to advancing the field. Meta-analyses that compare similar response variables and experimental conditions will produce the most meaningful results. When they are not similar and the differences are not specifically acknowledged, the analysis is lumping apples and oranges. This is a potential problem if it goes unrecognized, but a potential benefit if it allows you to generalize about many kinds of fruits.

A successful ecological study links the question and study design (chapters 1–3) with statistics (this chapter). Think through all three components as a group while you plan and modify your study. This chapter has addressed issues that apply to both manipulative experiments and observational studies. The next chapter offers some specifics about how to design and analyze observational studies that maximize your ability to learn about your system.

CHAPTER 5

Using Quantitative Observations to Explore Patterns

Earlier, we explained that manipulative experiments are our most powerful tools for establishing cause and effect in ecology (chapter 2). However, manipulative experiments are not always ethical, such as when asking questions about endangered or invasive species. In addition, they aren't always possible at relevant scales; for example, although some questions about reserve design can be answered with large-scale manipulative experiments, others may not be (see, for example, Newmark [1995, 1996] in chapter 3). Luckily, quantitative observations can also allow you to pose and evaluate hypotheses.

Observations of patterns form a continuum ranging from poking around to formal surveys. (By "surveys," we don't mean the questionnaires used by social scientists, but rather structured observations taken at different sites or at different times.) Surveys or quantitative observations are conducted in a systematic way. They allow you to observe patterns at the scales at which they naturally occur, and they can provide insights about relationships that you might not otherwise be able to evaluate. In addition, less work is required to survey a variable than is required to manipulate it. So, you can assess far more variables than you can manipulate, and you can observe patterns with less of an *a priori* understanding of the mechanisms producing those patterns. That said, the analysis of observational data is often less straightforward than that of manipulative experiments, and the inferences that can be drawn are generally not as strong. In the next section we provide some suggestions for developing, analyzing, and interpreting surveys.

Forming hypotheses from observations

Quantitative observations allow you to look at the relationships among many factors, and it can be tempting to record absolutely everything. But, as we mentioned in chapter 3, it is still important to start with clear questions and explicit hypotheses about the relationships or interactions before deciding on your study design. What do you ultimately want to know? We recommend that you start by writing down the causal paths that you want to consider (without worrying about what you can actually measure). These paths represent your hypotheses.

In the example involving deer and caribou shown in figure 1 (chapter 2), there were two different paths (hypotheses) that could lead to the negative effect of deer on caribou: deer could depress shared food levels (figure 1C) or increase levels of a shared parasite (figure 1D). Of course, a closer look at this system reveals many other factors that are potentially important for caribou, including anthropogenic factors such as logging and road

building, the predominant vegetation type, other competitors such as moose, and predators such as wolves
(figure 5, from Bowman et al. 2010). Ultimately, the goal
is to identify the factors that are important for caribou;
some of the paths are more likely than others, and some
can be eliminated completely. For example, logging can
affect the abundance of deer and moose by altering the
cover of deciduous trees (an important food source),
but deer and moose are unlikely to affect the amount of
logging.

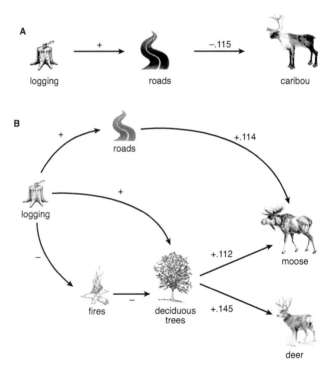

FIGURE 5. (*continued on next page*)

C

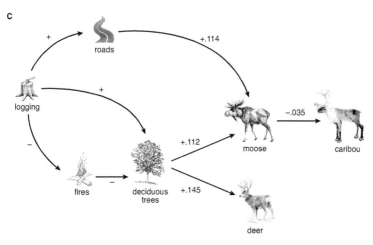

FIGURE 5. A path analysis based on results from Bowman et al. (2010) showing various hypotheses about effects of logging, roads, fire suppression, deciduous trees, and wolves, deer, and moose on numbers of caribou in northwestern Ontario (see figure 1 for some natural history of this system). The authors surveyed 575 cells (locations) of 100 km² each. (All numerical coefficients are significant.) A. Human activities, including logging and associated road building, negatively impacted caribou numbers. B. Logging and fire suppression had positive effects on the prevalence of deciduous trees, and deciduous trees were associated with increased populations of deer and moose. C. Roads were positively associated with moose, and more moose were associated with fewer caribou. As a whole, these anthropogenic activities increased numbers of deer and moose. Moose, in particular, were associated with fewer caribou (effect of deer was not significant and is not shown). D. These anthropogenic changes also increased the density of wolves, and wolves further decreased the number of caribou.

The path coefficients give an estimate of the strength of effects and allow us to calculate the relative direct and indirect effects of

(*continued on next page*)

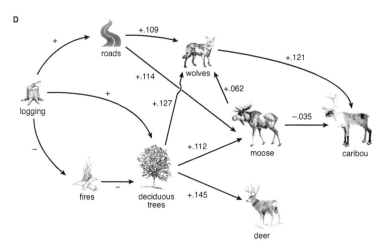

FIGURE 5. (*continued*) various ecological mechanisms. For instance, the
relationship between deciduous trees and caribou is the sum of effects of deciduous trees increasing moose, which decreased caribou
[(+.112)(−.035) =−.004], plus the effects of deciduous trees increasing
wolves, which in turn decreased caribou [(+.127)(−.121) =−.015], plus
the effects of deciduous trees on moose, which increased wolf populations, which decreased caribou [(+.112)(+.062)(−.121) =−.001].
Adding these three negative indirect effects together gives the sum
of the effects of deciduous trees on caribou, (−.004) + (−.015) +
(−.001) =−.020. It is apparent that the indirect path involving wolves
had the largest negative effect on caribou numbers. This analysis
suggests that the overall effects of roads and indirect effects mediated through wolf numbers were more important than effects mediated through potential competitors of caribou, that is, moose and
deer. This hypothesis could be tested by manipulating densities of
roads and wolves. This analysis omits other paths that may also be
important for caribou, some of which were included in the survey
but not found to be significant and some which were not considered (e.g., the meningeal worms mentioned in figure 1).

The scope of your study

One of the strengths of quantitative observations is that they allow you to observe nature at much larger spatial and temporal scales than manipulative experiments do. Since extrapolating beyond the scope of your data can be problematic, it is a good idea to conduct your survey at the scale you care about. For example, Walt Koenig and collaborators have been interested in explaining the boom-and-bust production of acorns by oak trees in California. They hypothesized that weather patterns might cause some of the year-to-year fluctuations that they observed at their initial field site in the Carmel Valley in California. At this field site, temperatures in spring predicted the size of the acorn crop (Koenig et al. 1996). Is this pattern limited to this one field site, or does it hold over the scale of interest, California? They expanded their survey and found a similar relationship between spring temperatures and acorn production throughout California (Koenig and Knops 2013). At their initial field site, trees were found to produce large acorn crops prior to years of heavy rainfall, a condition that favors seedling establishment, but this would suggest that oak trees predict future weather!?! However, this pattern did not hold at other sites in California and cannot be considered a general phenomenon (Koenig et al. 2010), much to the disappointment of weather forecasters. Surveys conducted at the scale that you care about are important in assessing the generality of patterns. In this case, surveys had to be conducted at multiple sites over the scale of the entire region since that is the scope over which they wanted to draw conclusions.

Surveys consist of measurements across space or time. Let's say we want to find out whether invasive plants are reducing pollinators for native plants. We could measure visitation by pollinators to native and invasive plants at multiple sites. This is called a cross-sectional or horizontal survey. Alternatively, we could conduct a survey of one location or with one native plant species, recording observations repeatedly over time. In this case, we care about changes *within* the site or plant species. For example, if we want to ask whether pollinator abundance changes as the proportion of invasive plants increases, then we could record pollinators over time at a site or sites where invasive plants are increasing. This type of survey is called a repeated-measures, panel, or longitudinal survey. A powerful survey design known as spatio-temporal analysis combines horizontal and longitudinal surveys; it allows us to compare patterns both between and within sites or individuals. If pollinators decline over time with increasing abundance of invasive plants within sites (longitudinal) *and* sites with more invasive plants have fewer pollinators (horizontal), you could make a pretty strong case that invasive plant abundance and pollinator abundance are causally related.

It's often hard to assess trends over time, especially for studies involving global change. Let's say we want to ask how rising temperatures will affect plankton communities in ponds. Because it is difficult to slowly raise temperatures in whole ponds, we may try to infer what will happen over time by comparing plankton communities in our study ponds with those in multiple ponds that are naturally warmer. That is, we substitute an effect of temperature variations over space (different ponds) for an effect we expect over time (rising temperatures due to climate change). Using natural variation over space (different ponds) to inform changes over time can introduce confounding factors. For example, if the warmer ponds we find are nearer the equator, they may have characteristics that tend to increase at lower latitudes (greater number of species, longer coevolution, stronger interspecies interactions). Space-for-time substitutions can be valuable, as long as you consider alternative factors that could produce the outcome you observe (Damgaard 2019).

Long-term surveys

Early in your career as an ecologist you are likely to feel pressure to produce results quickly, so it's easy to focus all your attention on surveys that can be completed in a season or two. It's worth keeping in mind that early in your career is also the best time to start collecting long-term data. Long-term data sets are valuable for expanding the scope of any experimental or observational study. If you have the opportunity to link your work to a long-term survey, it is worth considering.

Three keys to a successful long-term survey are (1) a clear hypothesis that the observations address (perhaps you see a theme here?); (2) stability of your site (will your site remain intact or turn into a shopping mall?); and (3) the simplicity of the measurements (will you feel motivated to conduct the survey while on crutches, toting a baby carrier, or facing other obstacles in your future?). Picking sites near your favorite vacation destinations or family members may also motivate you to make the effort over the long term. If you decide to start the world's newest 30-year data set, your much older self will likely thank you.

Factors to measure

The number of factors that you record for each replicate is also important. The simplest scenario is to record and relate two factors, where only one of the factors can affect the other. For example, you could relate the breeding time of a bird species with the average March temperature at multiple sites (replicates). It is reasonable to hypothesize that temperature affects bird breeding but not that bird breeding affects temperature.

This hypothesis could be complicated by adding more predictive factors (treatments). In addition to temperature, you may suspect that March precipitation will be important, and you may think that temperature and precipitation in other months could play a role in bird breeding as well. As the number of explanatory factors increases, so will the number of observations you'll need so that you can do statistics on the potential relationships between these factors and a response variable (in this case, bird breeding). At one extreme, it is impossible to relate more factors than you have independent observations; if you wish to explore 20 possible explanatory factors, you will need more than 21 independent observations (many more). Another situation arises when you are interested in the effect of a factor on more than one response. For example, you might hypothesize that March temperature (one factor) affects both the timing of bird breeding and the activity of snakes that eat bird eggs (two responses). You can undoubtedly think of more factors and response variables that could be interesting to measure. However, don't go around recording data willy-nilly; make sure that the factors you measure relate to your question.

There is some disagreement about whether it is better to design your survey to include only what is needed to test a single well-honed hypothesis, or to consider, in addition, other relationships that may not seem important from the outset. When we begin a survey, we attempt to have a clear, simple hypothesis in mind, but we also try to be flexible and opportunistic. If there are other factors that might be important and easy to measure, why not keep track of them? Of course, there are limits to what you can measure, and you don't want to dilute your efforts by attempting to keep track of factors that you will not use in your final model. For example, Rick has surveyed a population of wooly bear caterpillars at his field site for the past four decades (data are available at karban.wordpress.com/ltreb). Since they are heavily attacked by a tachinid parasitoid, he also recorded rates of parasitism by these flies, but parasitism has not proven to be a good predictor of caterpillar success or numbers (Karban and de Valpine 2010). More recent results indicated that ants and a virus can reduce numbers of caterpillars and that ants and rodents can be important predators of pupae (Karban et al. 2013, Grof-Tisza et al. 2015, Pepi et al. 2022). Rick sorely wishes he had recorded some quick-and-dirty estimate of ant, rodent, and virus populations during each of these years.

However, there are so many factors that could potentially affect numbers of wooly bears that it would have been prohibitive to keep track of all of them.

One way to decide whether to measure a variable in your survey is to think about whether it is likely to be an important component in your overall story. We may roughly divide the variables we measure in a survey into four categories, in order of importance: first, the response (the thing we're interested in explaining); second, key factors (the things we hypothesize cause the responses); third, interactors (things that change how the key factor affects the response); and fourth, covariates (things we want to account for, but do not really care about otherwise).

Measurements of your response and key factors are critical for telling your story. In contrast, we often get so hung up on measuring a zillion covariates that we lose track of the goals of our survey. A good way to avoid this is to imagine a news story about your findings. Does the newscaster say something like "After accounting for the effect of X, the researcher found that . . ."? If so, X is a covariate, and the things after the "found that" are the key factors and response that you need to focus most of your effort on. Interactors are similar to covariates, but more important because they affect the relationship between your key factors and response. If wetness of a site is an interactor, the newscaster might say something like "At arid sites, but not at wetter sites, the researcher found that. . . ." Covariates and interactors can be important because they factor out environmental noise and define the scope of your findings. But if they are not part of your primary hypothesis, they should not dominate the time that you spend recording data.

Conducting surveys requires less time, effort, and funding than setting up experimental treatments and provides the immediate gratification of putting meaningful numbers in your notebook. Surveys can also give you insights you cannot usually gain from experiments, and they can improve your story by adding realism and increasing spatial and temporal scale.

Causation and multiple factors in quantitative observations

We have emphasized that an advantage of manipulative experiments over observational surveys is the ability to infer causation with confidence. In many cases, there are multiple factors that could possibly cause a response, and it can be difficult to disentangle these potential drivers without an experiment. For example, recall the case in which March temperature could potentially affect the timing of bird breeding. Finding a positive correlation between temperature and bird breeding would not necessarily mean that the two are causally linked (figure 6A). They could be linked only through another causal factor (figure 6B), or not causally linked at all (figure 6C). If it were possible to experimentally manipulate the factors (temperature, insect emergence, day length) independently, we would be more confident about which ones cause which others.

Even when it is impossible to manipulate potential drivers, we may still be able to learn about causation. For instance,

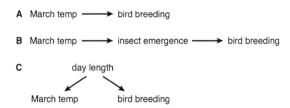

FIGURE 6. Possible relationships between March temperature and bird breeding. A. Variation in March temperature causes variation in bird breeding. B. March temperature causes a change in an intermediate factor, insect emergence, which then causes a change in bird breeding. C. A third factor, day length, causes changes in both March temperature and bird breeding, although temperature and bird breeding are not causally related.

Ian wanted to understand which leaf traits affected the survival of caterpillars that fed on oak leaves (Pearse 2011). It was difficult to manipulate many of the leaf traits in any meaningful way—how do you make leaves tougher or hairier? However, there was considerable natural variation in leaf traits among the oak species that he observed. So, he conducted a survey in which he measured multiple leaf traits of 27 species, placed caterpillars individually on leaves, and observed their survival. He assessed the relative importance of multiple factors (nine leaf traits) for predicting a single response (caterpillar survival).

In the simplest scenario, he could use regression analysis to separately assess the relationship between each leaf trait and survival. He could plot (or regress) each explanatory factor against survival and also examine the correlations among the leaf traits. Of the leaf traits he considered, evergreenness, leaf toughness, and tannins were the best predictors of caterpillar survival. This analysis of the observed variation provides some useful information about the traits that cause differences in survival. However, because he did not independently manipulate the predictor variables, many of the leaf traits are likely to covary, that is, be correlated. For example, Ian found that tougher leaves tended to have higher concentrations of tannins. Part of the statistical effect of toughness could actually have been caused by tannins or vice versa. If it was possible to independently manipulate leaf toughness and tannins, it would be possible more definitively to determine the effects of each factor on caterpillar survival.

When experiments aren't possible, there are statistical methods that can provide some inference about causality, and we will introduce several of these techniques, including multiple regression, principal components analysis, and structural equation modeling.

Multiple regression

When two or more predictors are highly correlated, it is difficult to distinguish the effect of one predictor from another. Multiple regression is a technique commonly used to divvy up the relative importance of collinear (correlated) effects of multiple predictor variables on a response. Multiple regression estimates how well the whole model (i.e., the combination of all predictors) describes the response and how well each predictor relates to the response after accounting for their collinearity with other predictors.

If two predictors are highly correlated, you may conclude that one is the causal factor when in fact the other is responsible for the effect. You can assess whether your predictor variables are correlated by plotting them. If they do not look correlated, there is likely no problem. If they appear to be correlated or you are not sure, you should calculate the variance inflation factor (VIF) for each predictor variable (Sokal and Rohlf 2012). VIF values greater than 10 indicate that predictors are too highly correlated to distinguish their separate effects and you should choose just one to use in the analysis (or use a technique like PCA, below).

Multiple regression assumes that the factors are related linearly, but ecological relationships are complicated, so this assumption won't always be valid. In these cases, more sophisticated statistical techniques may be useful, such as Poisson regression for count data, generalized additive models for complex linear patterns, and partial least squares when the number of sample replicates is small compared to the number of variables.

Principal components analysis

Another strategy is to look only at the shared portion (covariation) of the predictor variables to determine something about their overlapping effects on the response. Principal components analysis (PCA) condenses multiple factors that have the most covariation into a single axis. For example, when examining the effects of land use on eutrophication of lakes, several inputs (e.g., nitrogen, phosphorus, and salinity) may be collinear and these variables may serve as a condensed axis. In this case, these variables can be interpreted as surface runoff and this interpretation can suggest mitigation strategies. The downside of PCA is that it is often hard to interpret exactly what the condensed axis means in biological terms.

As ecologists gain access to big data, it becomes increasingly possible to examine the relationships among many factors. Models that include many different potential predictors will explain more of the pattern, but they may be more complicated than informative and be unlikely to represent a general pattern. In these cases, a challenge becomes separating those few meaningful factors from many others that might not be important. When you include many predictor variables, some might mask the effects of others even if there is not excessive collinearity between any pair.

Several information-criteria measures (including Akaike Information Criteria [AIC] and Bayesian Information Criteria [BIC]) have been developed to remove predictors that are not adding accuracy to predictions (Burnham and Anderson 2002). These techniques allow you to choose a relatively low number of predictors, making for a simple model that creates an accurate description of the response. A danger of this technique is that you may throw out a factor that is actually meaningful, or erroneously include a factor that appears significant by chance alone (recall the relationship between storks and babies, figure 4).

Structural equation modeling

Structural equation modeling is a technique that allows you to consider different paths (causal chains) and to evaluate which of those paths does the best job of explaining the observed patterns, i.e., which factors are likely to cause which other factors (Shipley 2000, Grace 2006).

Simple structural equation models are sometimes referred to as path analyses. The process of conducting an

SEM begins by sketching out the hypothesized causal relationships that link a set of variables. This is nearly identical to the process described in chapter 2 for forming
hypotheses based on observational data. The figures with
arrows that connected weather, plant stress, and herbivore
numbers are path diagrams. The paths you develop are
your informed opinions (hypotheses) about how your system works. The SEM that you calculate tests how well your
observations match with the informed opinion that you
drew out in your path diagram.

With a reasonable path diagram in hand, testing an SEM is
similar to tests of simpler hypotheses that we described earlier.
Numerous R packages, such a *lavaan* and *piecewiseSEM*, make
the implementation of an SEM relatively straightforward,
and tutorials exist to point the way (Grace 2006, Gana and
Broc 2019). There are a few practical considerations when
using SEM. First, this technique works best with complete
sets of observations. For example, if you wish to test the hypothesis that prairie dogs reduce vegetation height, which
then increases the abundance of grassland birds (Duchardt
et al. 2021), it is important to measure each of these three
variables for every sample. Second, because SEMs test complicated hypotheses, they require large sample sizes. A rule
of thumb is to conduct at least 10 independent observations
per variable in an SEM. So, in the example with prairie
dogs, vegetation, and birds, a minimum of 30 observations
would be recommended. Finally, SEM is increasingly flexible in accommodating data of various sorts; for example, you
can use measurements that are not normally distributed,
categorical variables, and variables coming from different

experiments or sets of observations. The utility of this technique to address complicated hypotheses that are commonplace in ecology will probably increase as we get used to analyzing our data in this way.

Interpreting "flawed data"

Many ecological studies are designed with inadequate replication and interspersion or other flaws in their design, analysis, and interpretation. If we identify such a problem in someone else's work, it is a natural temptation to disregard or trash the results, a phenomenon that Truman Young calls "pseudorigor." We suggest that this temptation should be held in check.

Even the best study is subject to alternative explanations because only a limited number of factors were considered. In chapter 2, we discussed Tom White's interpretation of the causal relationships between weather, plant stress, and outbreaks of psyllids. Since he only had a correlation, he couldn't establish cause and effect. Nevertheless, it would be a mistake to disregard his insights. Based on the information he had, he couldn't be sure of causality, but experiments done in the interim suggest that he was on the right track (Huberty and Denno 2004). Without long-term data and biologists who have (evidence-based) intuition about their organisms, we won't do the proper experiments, with or without statistical "rigor."

Too often people don't analyze data that they have collected because they feel their design wasn't perfect. Even if it isn't, the results may have something to teach you (and others). Even when data don't meet the assumptions required for inferential statistics, observations can suggest potential patterns.

Big data

Some ecologists never run an experiment or set up a survey of their own; instead, they synthesize patterns from increasingly large observation-based data sets that contain information relevant to ecologists, i.e., "big data." There are some questions that can only be asked at such a large scale that they would be nearly impossible to examine without synthesizing other people's findings. For example, testing whether patterns of plant invasions differ among continents (van Kleunen et al. 2015) would be difficult to accomplish without synthesizing multiple studies. Syntheses can ask about the generality of processes that were discovered at a smaller scale. For example, in greenhouse studies, garlic mustard reduced photosynthesis of neighboring plants by disrupting the symbiotic mycorrhizae of those neighbors (Hale et al. 2011). In a manipulative experiment removing garlic mustard, the neighbors that benefited were disproportionately plants that relied on mycorrhizae (Roche et al. 2021). A larger data synthesis allowed these workers to find that the abundance of garlic mustard was negatively correlated with the abundance of mycorrhizal plants over an entire region (Roche et al. forthcoming). In this case, data synthesis did not address a different question per se, but rather found evidence for an established pattern that held over larger spatial scales.

While big-data analyses open up an exciting new set of questions ecologists can ask, analyzing other people's data comes with limitations. First, it is a poor introduction to your system. If you start a project by setting up an experiment or a survey, you will likely form numerous alternative questions and hypotheses about your system by the time you're done with the season. However, the analysis of data may not be so inspirational, and your lack of personal experience with a system may lead you down dead ends or to unrealistic conclusions. Second, you cannot influence how a survey or experiment was conducted when analyzing other people's data, so it will be especially important to consider its limitations. Third, big data may offer a nearly unlimited sample size of observations. With such a large sample size, the distinction between statistical significance and biological significance is more likely to become an issue (see chapter 4). Finally, while analyzing other people's data would seem like an easy way to quickly conduct a project, you should not underestimate the time that it takes to wrangle data when it comes from other people and was collected for other purposes. Despite these limitations, big data can provide more generality and may allow you to see patterns at more realistic scales than single short-term studies at local sites.

In summary, surveys don't offer as much inference about causality, but they are generally better than manipulative experiments in terms of scale. Quantitative observations are often great ways to generate hypotheses, and statistical techniques such as structural equation modeling can provide some hints about causation from survey data.

CHAPTER 6

Brainstorming and
Other Indoor Skills

Unfortunately, field ecology doesn't take place exclusively in the field. In this section, we offer suggestions about techniques and skills that should help make your academic life a little better when you're back home. We'll cover how to brainstorm new ideas for your research, read more efficiently, organize and react to your field season (with a small foray into how to spend your time in the field), and approach writing throughout your career as an ecologist.

Brainstorming

Most scientists believe that research is a powerful way to learn about the world and that creative thinking is the engine moving research forward. Ironically, ecologists pay little attention to the research on how to think more creatively. We will attempt to summarize some of the large body of research on this topic, providing more references than in most other sections of this book. We will also outline a workshop that we have found useful for generating ideas.

A simple model developed by psychologists divides the creative process into two distinct tasks: generating ideas and evaluating them (Finke et al. 1992). Generating ideas requires expansive or abstract thinking, while evaluating ideas requires focused, concrete thinking. These two processes (generating and evaluating) are mediated by different neural networks (Ellamil et al. 2012, Jauk et al. 2012, Boot et al. 2017). That is, these processes involve different, often competing cognitive activities (Mayseless et al. 2014).

Research in this field has generally asked subjects (i.e., undergrads) to generate ideas in response to an open-ended prompt and then had trained raters evaluate the creativity of the ideas (defined as original and useful), standardizing for the reliability of different raters. This well-established method is not without issues but can nonetheless provide insights into ways to improve our ability to generate and evaluate ideas.

Research about generating creative ideas has produced two important takeaways. First, generating many ideas quickly can be an efficient path to creative ones (Paulus et al. 2011, Jung et al. 2015, Johnson and D'Lauro 2018). Longer lists of ideas, including bad ones, usually include better ideas than shorter lists do. For example, in one study, separate brainstorming groups were given 20 minutes to come up with responses to the same prompt. The number of responses varied greatly between groups (from 12 to 152 ideas). Those groups that generated more ideas (including bad ones) also generated the ideas of highest quality (Muñoz Adánez 2005). The mechanism is unclear. It may be a simple function of quantity (Simonton 2003). Alternatively, the most obvious or common ideas may be generated first and, when pressed to continue generating ideas, the mind may then make more remote connections, resulting in more original ideas (Kohn et al. 2011).

Our usual idea generation strategies (lab meetings and the Q&A period after talks) are oral group brainstorming sessions. People share ideas sequentially, and the loudest, most confident, or highest social-status members usually contribute the most whether or not their ideas are best. Sometimes discussions are channeled into a narrow range of ideas, and more interesting ideas may be left out. Oral group brainstorming does lead to new ideas, but it's not very efficient (Nijstad and Stroebe 2006). Written individual brainstorming followed by oral group discussion fosters more varied and higher-quality ideas than oral group brainstorming alone, when conducted with the same number of individuals for the same amount of time (e.g., Dugosh et al. 2000, Rietzschel et al. 2006, Girotra et al. 2010, Kohn et al. 2011).

The second takeaway is that you'll produce better ideas if you turn off your internal censor during idea generation. Our parents, teachers, friends, and society have all taught us to censor our thoughts and inclinations. Imagine what your social life would look like if you hadn't taken in their messages. But there's a downside to this socialization. Jeremy Bentham, the founder of utilitarianism, came up with an idea for social control in prisons that is more related to this issue than you might think. In the 1780s, he proposed a "panopticon" architectural design—a guard tower in the center and cells surrounding the tower, facing only inward (the Kyln in *Guardians of the Galaxy* employs this design). Guards in the tower would be able to see into each barred cell, but prisoners wouldn't be able to see into the tower's small windows. The panopticon, he argued, would keep the prisoners obedient because they would never know when they were being watched. The guards arguably don't even need to be there; fear makes the prisoners guard themselves. Foucault (1977) argued that the panopticon is emblematic of our society, including academia. We internalize the guards' power over us and police ourselves. We censor our own ideas before they even form. So, to generate new ideas, we need to temporarily block out the guards in our heads. If we explore our "bad" ideas for a while, we might find that much better ideas emerge from them, directly or indirectly.

Even if you learn to turn off your own censor when appropriate, you are still working among other people, some of whom will want to play the role of guard. Academics can be competitive and judgmental. Fear of ridicule can reduce our ability to generate ideas because doing so becomes risky (Janes and Olson 2010, 2015). It's fine to joke around, just not at others' expense. Environments where people either don't use ridicule at all or ridicule only themselves have both been found to encourage risk taking and creative thinking (Janes and Olson 2010, 2015). Find yourself a group of friends and advisors who welcome uncompetitive conversations and foster risk-taking (Edmondson and Lei 2014), the "risk" being ill-formed ideas that may come out of your mouth in front of them. You're likely to end up with much better ideas while hanging out with them than with people who joke about each other's comments.

Techniques such as deliberately shifting perspectives and working within constraints can help generate ideas (Torrance and Torrance 1978, Rosso 2014, Marguc et al. 2015, Runco 2020). Mikaela has applied these research findings and techniques to a workshop we have used at multiple universities. If you want to think up potential research questions, alternative hypotheses, or even career ideas, you can run the workshop yourself and come away with lots of new possibilities (see box 4).

Evaluating creative ideas is in many ways more difficult than generating them. Recent studies have found that we are not good at recognizing or choosing the best ideas, defined as those that optimize originality and usefulness (e.g., Rietzschel et al. 2010, Johnson and D'Lauro 2018, Zhu et al. 2020). In fact, studies keep turning up the discouraging result that we are no more likely to choose the best idea than if we picked one at random. There are probably lots of reasons for this. People are biased toward ideas with which they have past experience (Eidelman et al. 2009). People who face an uncertain future or are in a high-stakes situation (two scenarios that academics experience a lot) are less able to recognize creative ideas (Mueller et al. 2012). Experts in a field may choose less creative ideas because they want (consciously or unconsciously) to maintain the status quo if it is linked to their social status or self-worth (Cropley 2006). Keep that in mind next time an "expert" dismisses the ideas of a more junior person. And pay attention to your own biases when sifting through your many ideas. Don't dismiss an idea just because it surprises you.

Box 4. *Generating ideas*

This workshop is designed for a group of people to help one focal person generate ideas for a research question, alternative hypotheses, or whatever. It can also be modified so that each participant works alone on their own question. (This second format allows people to address personal questions, such as what matters to them in their career or how to find work-life balance.)

Before the workshop
Select one focal researcher and six to eight other participants who vary in their familiarity with the researcher's question and system. Ask the focal researcher to prepare no more than eight minutes of background information and preliminary data for their topic, ideally as a handout or PDF they send to all participants for reference. Set aside at least two hours.

On the day of the workshop
Bring fine-line markers (not Sharpies—they're too smelly) for everyone and at least 100 colorful 3″ × 3″ sticky notes or small squares of paper and tape. Provide something to drink and pizza or a snack. The goal is to create a playful, friendly atmosphere.

Pre-brainstorming

Workshop instructions
The hardest part about running this workshop is that people want to do it out loud. Set different expectations beforehand. Explain that research has demonstrated that individual, silent brainstorming is more efficient than oral, group brainstorming (see text). Tell them that you

Box 4. Continued

will give them a few opportunities to talk but that most of the activity will be quiet. Explain that they will generate ideas silently and record each idea (no matter how half-baked) on a separate sticky note or paper square. Keep the tone friendly; despite the workshop being silent, the mood should be upbeat (Baas et al. 2008).

Brainstorming instructions
Give a description of the four "rules" for the workshop (Osborn 1953):

· Aim for quantity, not quality, of ideas.
· Withhold criticism.
· Combine and improve ideas.
· Encourage wild ideas.

Each idea—including the "bad" ones—that a participant comes up with should be summarized in a few words or illustrations on a sticky note.

Focal research overview
Next, the focal researcher briefly describes the phenomenon, observation, or challenge that they would like to examine creatively. The focal researcher or the participants often push to lengthen this period, not realizing that the most productive components of the workshop will happen later. The focal researcher should end with an explicit prompt for the group, such as "Which research questions can I ask to build on my current information?" or "Which alternative hypotheses should I consider?"

   Let participants ask (quick) follow-up questions about anything they don't understand. If someone stops asking follow-up questions and instead starts offering ideas, kindly ask everyone to record their ideas on sticky notes, and then move to the next step. Most of the time, doing this early (that is, redirecting talkers to the sticky notes) creates a culture of silent idea generation that last the rest of the workshop.

Brainstorming

The rest of the session is a series of "rounds" (prompts) to generate as many ideas on individual sticky notes as possible. Allot about 10 minutes for the first round and 20 minutes for each of the other rounds. After each round of brainstorming, everyone sticks their notes on the wall and looks at others' notes, gallery style. After a few minutes of gallery viewing, if time allows, encourage participants to (briefly) say an idea or two that they are most excited about and to (again, briefly) build on the ideas of others. Then quickly move on to the next round. Plan for a few minutes for the brief check-in and transition between rounds.

Round 1. Initial ideas

Ask each person to write their initial ideas on sticky notes. These are usually ideas that they are retrieving from memory (Silvia et al. 2017) and may limit their ability to see other ideas (Crilly 2015).

Round 2. Points of view

Ask group members to haphazardly call out people, organisms, or anything else that could have a point of view that are unrelated to the research topic (such as *a citizen of Wakanda, a solar flare, your aunt*). The more surprising, the better. Write each of these points of view on a whiteboard until you have about 20. Alternatively, select people or organisms suggested by a random generator, such as ideagenerator .creativitygames.net. Then ask participants to choose an unfamiliar point of view from the list and use it to answer the researcher's question. If they run out of ideas from that point

Box 4. Continued

of view, they should move to a new point of view until time is up. Try to use approximately four points of view.

Round 3. Constraints

Now ask group members to haphazardly call out uncommon constraints starting with "The study must . . ." (such as . . . *use olfactory information,* . . . *be conducted without electricity,* . . . *involve the color red*). Again, the odder the better. You may choose to use ideagenerator.creativitygames.net to spur ideas for unusual constraints. Ask participants to silently write ideas based on one or more of these constraints. Again, remind them to move to a new constraint whenever they are ready.

Round 4. Paired discussion

Ask participants to pair up so each one can discuss one or two ideas they find particularly interesting. Participants normally choose to walk around outside during this round (see Oppezzo and Schwartz 2014). When they return, they should record each of their new ideas on sticky notes.

Round 5. Round robin

Another way to keep novelty high is to give people a chance to quickly swap ideas (Coelho and Vieira 2018). Place multiple blank sticky notes on individual sheets of 8.5"×11" or A4 paper, and give one sheet to each participant. Ask participants to survey the sticky notes on the wall, copy one particularly intriguing idea onto their first sticky note, and riff off of that idea on the second sticky note. Each sheet is then passed to a different participant, who uses the previous sticky notes as inspiration for a new idea. Each page of sticky notes gets passed to each participant and eventually returns to the first participant, who can add a final idea.

Round 6. Implementation criteria
By now, this process will have generated a range of ideas: from mundane to original, from straightforward to impossible to implement. Ask each participant to choose the idea they think is most original, even if it is currently impossible to implement. Tell them to generate methods for how to implement the idea and record them on sticky notes.

Collect the ideas
At the end of the session, the focal researcher receives the sticky notes and sheets of paper.

After the workshop
Within a few days of the workshop, the focal researcher should collate and annotate the ideas. It is useful to group the ideas and mark which ones are most unusual, exciting, likely to contribute to the field, and so on. To select ideas, they should ask "Why is this question important?" In one study, subjects who asked *why* rather than *how* (e.g., "How can this question be asked?") chose more creative ideas (Mueller et al. 2014). Finally, they should collate the ideas and develop a preliminary follow-up plan. Checking in with an agreed-upon "accountability partner" afterward is a great motivator for this step.

The goal of all researchers is to push the field forward; doing that requires new ideas. Many of your ideas will come spontaneously. But you can also generate ideas deliberately, without waiting to "feel inspired," to make sure you aren't missing the most exciting ones. Also, generating ideas is fun and a change of pace from the rest of your work.

Reading like a scientist

By the time you get this far in your education, can we really have anything to say to you about reading that you don't already know? Reading the academic literature is not the same as reading novels, blogs, or textbooks, and what works as an undergrad won't necessarily work in grad school and beyond. Knowing the structure of journal articles (described in the section about writing journal articles in chapter 8) makes it easier to focus on specific parts of a paper if you want to find out where the field is currently or what specific questions were addressed by the authors.

One problem that all ecologists (new and old) face is the overwhelming quantity of information available. It is not possible to read all of the relevant literature, but letting yourself become overwhelmed and then reading little is not a good strategy. The key is focusing on the topics that are interesting to you, especially on the important papers about those topics. Review articles often provide an excellent entree into a field. Look for papers with "review" in their titles or those published in one of the *Annual Review* or *Trends* journals. The introductions to research papers generally cite the important papers in a field, providing a starting place. You can also enter keywords into the Web of Science database to gain a foothold, or use this database to see which other articles have cited an article that you find particularly relevant. Don't get distracted; most of these won't be helpful enough to pursue in your limited time. You don't have to cite every last article on your topic. Your major professor or other mentors may also point you to important papers. Even when you invest in an article, skim for meaning only.

You may want to consider how you read. First-year law students have been the subjects of research on reading strategies that may also apply to ecologists. The most common reading strategies have been categorized as "default" (summarizing, paraphrasing, underlining), "rhetorical" (contextualizing, connecting with prior knowledge), and "problematizing" (raising questions about the meaning of the text, posing problems, solving problems; Deegan 1995). Students in the bottom quartile of their class were most likely to rely on "default" strategies, which probably won't work for grad students either.

The students in the top quartile of their class tended to use the "problematizing" strategy, which is probably the most valuable reading strategy for building a successful research project. Before reading a paper, decide specifically what you want to get from it. Do you want new ideas for research directions? Do you want to know what other people have been working on in this area? Are you interested in a technique that the authors used? Give yourself a time limit to get the information that you are looking for—get in, get what you want, get out. Allow yourself to enjoy reading within these boundaries. How do the authors justify their work? What are their main results? How do they relate their work to the bigger picture? You can skim for answers to these questions for a few minutes, but keep in mind what you came for.

Write a few sentences in response to the question you wanted to answer before you started reading. Rick likes to put the question at the top of the article and then to note the page or paragraph where the authors addressed his specific concerns so that he can return to the spot. This technique keeps you from using reading to procrastinate, and forces you to deal with the material and to write. This strategy will help you make sense of your reading now and will make the writing process easier as you tackle it later.

Many grad students get stuck looking for the flaws in papers. Every paper has them; finding flaws doesn't necessarily justify discarding the work, and it may distract from the useful contribution that the paper can make. In general, resist the temptation to be a critic rather than a producer.

The problematizing strategy above may also be helpful while listening to seminars. Thinking about how the research relates to your work or preparing yourself to ask a question (whether or not you actually ask it) can make you a more attentive and less passive listener.

You have been reading for quite a while, and it may take some time to train yourself to read in a new, more efficient way. We hope this section encourages you to analyze your own reading strategy and talk to your colleagues about strategies to get through the mountain of literature that we are all facing.

Organizing and reacting to a field season

Organizing a field season requires thinking of both the big picture and the day-to-day. Remember that you will want to share your results as a story. So, ask yourself throughout whether your questions (the big picture) and your research activities (the day-to-day) are allowing you to fully develop the story your system is offering. Actually write out a plan for your season, don't just plan it in your head.

Before you go to the field

Begin by making sure you have a firm handle on your questions and that answering them is feasible. Write down your questions. Next, consider the manipulations or observations you can conduct to test your questions. It is amazing how often the manipulations and observations that we conduct don't answer the question that we want to address. This disconnect can often be avoided by being very explicit about your question and then asking whether the measurements you plan really are the most appropriate match for that question. Consider whether alternative methods and measurements could answer your question more effectively. Run your ideas past your major professor, lab mates, or anyone else who will give you meaningful feedback on your plans. Then do some more writing before you forget their comments.

Make sure that your plan is feasible—that you have access to your field sites, necessary immunizations or safety equipment, money for transportation and supplies, that you are not proposing techniques you can't possibly do, and so on. Try to anticipate all the things that could go wrong in your experiments or observations and how you will respond to them. Troubleshoot them as much as you can, and come up with contingency plans. What will happen if your organisms are harder to find than you expect?

What kind of weather could interfere, and what will you do about it? Again, run these potential problems and solutions by someone who knows the system (your major professor or a colleague) to make sure that your ideas are on track. Picture all the steps that you will go through. What will you need in order to accomplish these steps? Don't blame yourself if things go wrong. No matter how much planning and troubleshooting you do, equipment will fail, organisms won't behave, or environmental conditions will be uncooperative. Just keep moving forward with whatever does work.

Before you go out to the field, prepare your data sheets. Include all the variables that you plan to measure as columns and each of your independent replicates as rows. Include columns for replicate number, treatment, and block if your design has these variables. If you are new to ecology, get feedback on your data sheets. After your first trial run, your data sheets may need to be revised.

While you are in the field

Once you start your fieldwork, let your observations and results direct your next steps. Go where your system wants to take you. Constantly reassess what you can do and what alternative explanations and paths could be. Take stock of your progress and your preliminary results at least once a week. Informal writing can help with this process.

Many of the best scientists are both very lucky, and perceptive enough to recognize that they have stumbled upon something unexpected and novel. Keep your eyes open and be willing to take off in a slightly different direction if a valuable, unanticipated observation presents itself. Go in the directions that you determine will be the most profitable, and don't feel constrained to follow the plan that you set for yourself before you started. Louis Pasteur is quoted as saying that in the field of observation, chance favors the prepared mind. Reading broadly and going to talks help develop a prepared mind. That said, there is no substitute for getting out and poking around.

Be nimble about adjusting your plan. The historical development of your thoughts will likely be interesting to you personally but shouldn't be included in your eventual story. As you assess and redesign your experiments, focus on developing your current story rather than pursuing the one you started with.

Document your day-to-day work by writing down and thinking about each day's activities, observations, and data. After setting up each study, take careful notes on your methods; if you don't, you run the risk of forgetting key details you'll need later. Take photos of your system to document what you've done (and of anyone who helps with your study). You can use these when you present your story later. Throughout your field season, keep tabs on your results and what they mean. Do descriptive stats, such as calculating treatment means and variances. This will help you get a feel for your data and their patterns, whether your sample size will be sufficient, whether you are missing a key observation or experiment, and so on. In short, make sure you process what your field season means even while you are in the middle of it.

At the end of your field season

Back at home, do a complete statistical analysis and write up your results as a preliminary paper even if you know that more fieldwork is required. (See "Journal Articles" in chapter 8, for advice.) Analyzing your data lets you determine which effects are "real," which experiments you want to repeat, and which new directions you want to think about. The reading you do to assemble the introduction and discussion can be particularly valuable as well. Extracting the relevant concepts from other studies lets you put your questions and results in a broader context and gets you familiar with the pertinent literature. (Note to perfectionists: Don't get hung up here. It is more important to get a preliminary draft of your results than to do a complete literature review.)

Once you have a draft of your preliminary results, ask for feedback from your colleagues and committee members. They will help you see things you missed and may have literature suggestions. Use their feedback to revise your paper and to fine-tune your plans for your next field season.

Writing a preliminary draft may seem like extra work. It's not. You can use the draft as a framework for your final paper. It also lets you see what you have and what you don't, which will make your next field season more efficient.

Writing habits

Writing is difficult for many people, but we have learned a few tricks that can make it easier (again, see chapter 8). Because research involves producing a final written product, people sometimes assume it means you shouldn't write up your research until the end of your degree. We don't recommend waiting that long. Instead, write early and often. For example, it can be effective to write every day for a brief period—say, a half an hour, or even less. The amount of time isn't important; the daily routine is. Start your writing routine years before you are ready to write up your thesis. Write to process papers you have read, document your methods, generate ideas, and keep track of your observations. Charles Darwin wrote every morning and was done by lunchtime. This habit allowed him to produce about 20 books and monographs.

Another trick we have found very useful is called the Pomodoro Technique®. (Google it if you're interested.) The Pomodoro Technique® is a (ingeniously?) registered trademark by Francesco Cirillo, who developed this method. He advocated working for 25-minute intervals, punctuated by five-minute breaks. At the time, most kitchen timers in Italy, where Cirillo lived, were shaped like tomatoes (*pomodori*), which inspired the catchy name. You can experiment with the duration—a popular variant is 50 minutes of writing with a 10-minute break. We have occasionally been invited to do this technique with groups of grad students at UC Davis and found it motivating and fun. It may be useful to briefly discuss your goals for the session before starting.

One of the best things about being a field ecologist is being in the field. But when you're not (which for many ecologists is the majority of the year), there is still work you can do to make the time productive and fun. We hope this chapter has given you useful techniques for brainstorming great ideas, reading and writing more efficiently, and planning to get back in the field again.

CHAPTER 7

Working with People and Getting a Job in Ecology

When we aspired to become biologists, we imagined that the scientific process was totally objective, and that the truth could be separated from lesser hypotheses in a manner that was removed from social interactions. The longer we stay in this business, the more we are struck by the opposite. Science is a social endeavor, most ecologists are human beings, and coming up with good ideas isn't enough; successful ecologists need to be persuasive about the value of their ideas.

In this chapter we will consider some of the social situations that you are likely to encounter as a graduate student and as a professional ecologist. We'll cover picking a professor and a committee, interacting with other ecologists, and applying for jobs after grad school, among other things.

The model for academia was established during the European Middle Ages, and institutions of higher learning around the world still cling unconsciously to medieval European conventions. These include what Damrosch (1995:18) described as "the indentured servitude of graduate student apprentices and postdoctoral journeymen." Perhaps because universities originated from monastic traditions, a high degree of zealous dedication and self-discipline is expected. Any less is likely to be met with disapproval. Many administrators at universities talk about "work-life balance," but the academics that you encounter (your major professor, search committee members, etc.) may have no such balance in mind.

Things to know before applying to grad school

The prestige of the university granting your degree is far less important than the level of intellectual, emotional, and financial support that you receive along the way. You should try to get a funding package that will keep you from needing outside employment, but you probably shouldn't pick a program based on a few thousand dollars more or less. Grad school is a mistake if your goal is to get rich; you can increase your income by doing just about anything else. Entering a graduate program means deferring for many years earning enough salary to buy a house or put away money for the future. But grad school may be a great place to be if you are passionate about ecology.

When you are choosing a graduate program, pick a professor rather than a university. You should contact potential major professors (referred to as advisors at some universities) when applying to programs. How do you find potential major professors to contact in the first place? One effective way is to see who has done work that excites you. Reading broadly can help you discover areas that you find particularly cool. Google Scholar and Web of Science can link you to other people who have asked similar questions. You may also be able to get useful suggestions by talking to instructors of the undergraduate classes that you most enjoyed.

Undergrads and recent grads often find many research areas fascinating and aren't sure of a direction. It may be useful to gain experience and hone your interests by working as a field assistant. One good source of opportunities in diverse ecological fields can be found at the Wildlife and Fisheries Sciences Job Board maintained by Texas A&M (https://wfscjobs.tamu.edu/job-board; search for "field assistant" or "field technician").

When it comes time to contact potential professors, write a formal email rather than the sort of message that you might send to a friend. You can google samples of academic cover letters for graduate school. Resist the temptation to explain that you've always been interested in nature and that playing in the mud has been your passion since the age of three. Instead, be prepared to explain why working with that major professor will be a good match for your current interests. This person will be more impressed if you discuss their actual research than if you say something generic about wanting to study charismatic vertebrates or work in the intertidal. Similarly, when it comes time to submit a statement of purpose with your application, it should focus solely on your research interests.

Make sure that you can communicate well with your major professor. The reputation that your major professor has as a researcher is important but not nearly as important as their reputation as a mentor of graduate students.

Visit potential professors before accepting any offer. Do your personalities seem like a good fit? You will be surprisingly dependent on this person for advice, financial and emotional support, and help getting grants and a job. Does their vision match yours? Find out how long your leash will be. Are students expected to put in face time from eight to five, or can you set the schedule that allows you to be most productive? Will you be given a project, or will you come up with your own? Ask about what has happened to previous students. How many have finished? If they didn't, why didn't they? What jobs did graduates get? Ask current students privately what it's like to work with that major professor. Students who are miserable will probably be kind enough to warn you. It is extremely helpful to correspond with past students. Would you like to be in their shoes five or ten years down the road? Professors have track records, and you can expect to face many of the same situations and eventual successes as past students. If you are not yet a grad student, you will probably underestimate the importance of the relationship with your major professor. However, it will color everything that happens to you, so choose carefully.

One subject that grad students and major professors often butt heads about is authorship. Each subdiscipline has a slightly different tradition about what constitutes authorship, with lab-intensive programs including the major professor as an author more often than field-oriented programs. It's a good idea to ask former and current students about joint publications with your potential future major professor. The Ecological Society of America has

developed a policy statement for authorship (https://www
.esa.org/about/code-of-ethics/). According to this policy,
to be considered an author, a person must make substantial
contributions. Simply providing funding does not confer
authorship privileges. Even if it is awkward, discuss norms
and expectations about authorship (including author order)
when meeting with potential major professors. You may
want to keep this fluid, but having this discussion early on
can reduce the chances of big disappointments later.

Interacting with other ecologists and beyond

Despite the myth, grad school and ecology in general are
not completely solitary pursuits. Your work as a grad student includes exams and a thesis, both of which involve
committees of professors. You will be collaborating on research, attending conferences, and enlisting the help of a
wide range of people in setting up lab and fieldwork. Success in ecology will ultimately depend on these interactions with other people.

Completing grad school often requires assembling at
least two committees—one for your qualifying exam (QEs
or orals) and the other to assess your thesis or dissertation.
Involve your committee members so that they can provide
advice and support much as your major professor does.
Choose committee members who are going to give you the
most help. Don't be shy about asking for help—mentoring
grad students is part of their job, and many professors find
it really satisfying. Get to know them by taking their seminars and talking to them about your project. We found

attending the lab meetings of our committee members extremely helpful in our development. Try to use the same people for your exams and thesis committee so that you know each other well. The better they know you and the more invested they feel in you, the more help they are likely to give you when it comes time to submit your manuscripts for publication and apply for jobs. Don't hesitate to run your research proposals, grants, and manuscripts past your committee members. If they are too busy, they'll tell you, but at least they will know that you are trying hard. If they are always too busy, think about finding committee members who will invest more time in you. Don't make the mistake of "getting lost" interacting only with your own lab—you will need relationships with other professors to succeed.

If you intend to become an ecologist at a research-oriented university, your committee members are also likely to be a good source of advice about career ideas and application materials. If you plan to seek a job elsewhere (a small liberal arts college, governmental organization, or NGO), these same committee members may or may not "get" your vision for your future, depending on their own biases about jobs outside of research universities. Also, even those who have the best of intentions usually will not know as much about the culture of your preferred career path as people who are already on that path. Some grad programs will allow you to add someone to your committee who is from outside your university; this may be a good choice if you have a clear idea of the type of institution you will apply to when you finish.

Both as a grad student and as a professional ecologist, you can expand the scope of your research by collaborating with other scientists. Good collaborations, like other mutualistic interactions, generally involve organisms that can offer skills or expertise that the other partner cannot easily acquire on their own. Perhaps one person is a good chemist, while another has little expertise in chemistry but knows a lot of natural history. Together this team is capable of going places that neither person could go alone. Sometimes people who have similar interests but different personalities are able to work together very effectively; an idea person who has trouble finishing projects can work well with a pragmatist who is a little less creative. The main disadvantage of collaborating is that you lose some control over the content and pace of the work. For example, you may be in more (or less) of a hurry than your collaborator, which can be frustrating to both of you. Despite this constraint, the advantages of collaborating are great, and the rugged individualist who works completely independently is becoming less common in our field. In the past, most papers in ecology were single-authored; today very few are. When you collaborate, make sure to discuss authorship and author order early in the process. This can prevent tension and frustration later.

Another valuable way to interact with colleagues is by attending meetings and conferences. Even if you find these events stressful, it is worth making yourself go and trying to interact as best you can. Be kind to yourself at meetings by not having unreasonable expectations, and by taking breaks (naps, walks) when you are feeling fried.

Conferences give you the opportunity to let other people know what you are doing, to learn what other people are up to, to get feedback, and most importantly, to schmooze. Don't feel like you need to introduce yourself to the headliners in your field. Any and all contacts can be beneficial. Getting to know the people in your field personally is extremely worthwhile and will help you start collaborations, get your manuscripts and grants accepted, and let you feel part of a community.

Social media can be another way to interact with people who share your interests. Networking of all sorts is generally useful, and social media has the potential to let you develop relationships with a much more diverse, far-flung group of people than you might otherwise reach. Different platforms serve different purposes. Think about what you want to achieve and how best to achieve it before investing much energy in a new platform. As you might guess, in order to connect with people who share your interests, follow or like their social media accounts and content, and post relevant content to your own accounts (with high-quality images to grab attention). If you don't have your own results or publications yet, amplify content from other researchers by linking and retweeting.

Be mindful about your use of social media. Keep the content associated with your real name professional. Almost all potential employers will google your name; make sure that what they find is your Google Scholar or LinkedIn profile, not photos of you at keggers. Stay aware of the ways social media can drain your mental reserves; for many people, it can be addictive or discouraging. If you decide you want to invest in it, remember to prioritize your research so that you can finish your degree.

Interacting with people who aren't professors or grad students

Research projects and careers in ecology put you in contact with a wide diversity of people, including administrative assistants, reserve stewards, and resource managers, among others. They often have specialized skills and knowledge, and they definitely have lots of work of their own to do. Unfortunately, some ecologists see the world as a caste system, with academics on top. The good relationships you cultivate with the people who facilitate your work can be mutually rewarding. Even if you are shy, make an effort to communicate your respect and appreciation to everyone who facilitates your work.

One way to get research accomplished is by hiring helpers, usually undergraduates. They can do repetitive work, freeing you to do more creative tasks. Research assistants can also get work done at your field site when you can't be there. In addition, some projects require a lot of hands; assistants can allow you to answer questions that you could not address as one person. If you decide to hire assistants, remember to acknowledge them by name in papers and at conferences.

We have found that there can be severe downsides to hiring helpers. Hiring assistants is expensive, and they have much less invested in the quality of the work than you do. For many assistants, it's just a job. Assistants generally don't have the expertise and intuition for the system that you do.

Often when we have asked other people to do routine tasks, we have been surprised when the tasks weren't done "right." There are many little things, "tricks" if you will, that each of us take for granted. It is very difficult to convey all of these to another person, no matter how detailed the instructions. One way to minimize the risks is to work along-side your helpers. That way you can provide more quality control (and, if your assistants plan to apply to grad school, mentoring) and also develop your own critical intuition about your system. Another issue is that hiring helpers generally involves you in work like writing grants and progress reports, and this paperwork can make you the administrator and your assistants the ecologists. You may miss out on developing intuition about the details of your system (and maybe about the big questions as well). In short, make sure hiring assistants really will help you accomplish your goals.

Getting a job in ecology

Getting a job is one of the most important things you'll do, and something that you'll want to start thinking about years before you finish your degree. Think carefully about the kinds of job that you would like, how to find them, and how to make your application attractive.

Begin by figuring out what type or types of job you are after. Since no job will offer everything, it is worth spending some time thinking about what will matter the most to you. As a (very) general rule, graduate degree–level jobs in ecology offer a decent amount of freedom to pursue your own interests and a self-driven schedule, though they don't offer much geographic flexibility. If you want a job in a specific place, you may find that you need to compromise on freedom in your research pursuits or your daily schedule. Additional attributes will make a big difference in your work life too. Do you want job security? A chance to teach or mentor? A certain salary? Before starting your job search, ask yourself which attributes you want to prioritize. This question may be difficult to answer: How can you know what jobs might be available and what it might be like to do them? You may have been trained (brainwashed) to value some jobs over others because of the prestige they will bring.

We recommend Beck (2001) for an in-depth and surprising treatment of this subject. While it is not research-based, it is full of wise insights. Beck distinguishes the "social self" from the "essential self." Your social self values the things that are valued by the people around you. Your essential self knows how you truly want to spend your time because it has "compasses pointed towards your North Star" (Beck 2001:3), but it may be repressed by your social self. Many of us subconsciously allow other people to direct our choices away from the paths we would otherwise most enjoy.

It can be more difficult to untangle the influences (people) that direct your path than you might think. For example, Mikaela entered graduate school with the express intent of eventually working at a small, teaching-focused college. Within a few months, she'd forgotten about that and "realized" she wanted to be a research professor. It took several misdirected years before she even questioned where this new aspiration had come from. (She eventually noticed that she'd been hypnotized by the culture of R1 institutions, and she got a job helping academics improve the way they teach science, which is a much better fit for her.)

Who are these people who are squelching our internal desires? They are people who, in most cases, have the best intentions for us. Disappointing these people can be very painful, but allowing them to direct your path has grave, and often hidden, consequences as well. Beck has many exercises to help you figure out what your essential self wants and how to get there.

Once your goals are clear to you, it is important to be strategic to accomplish them. As we mentioned in the Introduction, the most heavily weighted currency for essentially all jobs in ecology (academic or otherwise) is publications. (Jobs at community colleges may be the one exception.) Whether or not the job will require research, search committees will evaluate your application by looking at your publication record. Skills, support, and collaborations that will help you build that record are worth pursuing.

As you become clear about what you want to do, make sure you find out about the particular subculture of the "club" you hope to join. Let's say, for example, that your focus is community colleges. Community-college professors have told us that grad students who teach a community-college course while still in grad school are more viable candidates for community college tenure-track positions than grad students who don't. Waiting to get community college experience until after finishing your PhD puts you at higher risk of becoming a permanent adjunct professor or "freeway flyer" who gets hired for single courses at multiple schools but isn't seen as tenure-track material, regardless of your skill set. (Cash-strapped administrators have figured out that they can hire freeway flyers for much less money than tenure-track faculty, even if the arrangement comes at the expense of the adjuncts, students, and the academic community.) Each subculture has its own unpublished rules, so it's worth asking around.

Potential employers focus heavily on your ability to do and publish research, but they expect other skills as well. For example, many nonacademic jobs—especially nonprofits—require interpersonal, leadership, networking, and management skills (Blickley et al. 2013). Private-sector search committees may be interested in technical and technological skills. Teaching-heavy positions will require communication and organizational skills as well as knowledge of how students learn. You get the idea.

Often, you are building those additional skills whether you mean to or not. Being a teaching assistant demands skills that are transferable to many jobs, not just teaching positions. For example, grading papers may help you acquire skills that will be useful later, when you need to give feedback to people you supervise. Being a TA in a lab class gives you the opportunity to practice presenting in front of an audience, explaining difficult concepts clearly, and making decisions with colleagues (other TAs). When you put your job application and cover letter together, describe how the skills you developed in grad school "transfer" to the job requirements.

Jeremy Fox has analyzed the success of ecologists who are seeking tenure-track jobs in North America (Fox 2020). He found that at least a third of PhDs who got as far as the application process were able to land tenure-track positions. Said another way, up to two-thirds of highly skilled, highly trained PhDs who wanted academic positions didn't get them. The ability to conduct fieldwork was the most commonly listed qualification, appearing in nearly two-thirds of job ads. Jobs at R1 research universities received more applicants than other tenure-track jobs. People who submitted more job applications reported getting more phone, video, and campus interviews, although the benefits of more applications appeared to plateau at around 30 per year. Perhaps the most valuable personal attribute for ecologists is persistence. Most assistant professors had several years of post-PhD experience under their belts before getting hired. Few applicants were able to leave academia and then return to fill assistant professorships. This last finding backs up an unwritten rule in the academic job market: it's rare to leave "temporarily" and still be considered hirable as an academic.

When you are ready to apply for jobs, and probably long before, check out the listings on the Ecological Society of America's online job board (https://www.esacareercenter .org) and at Ecolog-L. Two other outstanding resources for academic jobs in ecology are ecophys-jobs.org/positions .html and ecoevojobs.net. It's worth asking people who have jobs in your chosen field if there are websites specific to your interests.

Many of the most satisfying jobs are never listed—you create them. Figure out what skills or expertise you have that would be valuable to an employer. Next, determine who might find those skills or experiences appealing and might be in a position to offer you a job. Approach that person with your proposal and make the case that you can provide something that they will value. The popular job-hunting book *What Color Is Your Parachute* (Bolles and Brooks 2020) has advocated this method of creating your own job for decades. If you can get past the silly sketches and exercises reminiscent of middle-school workbooks, the book can be remarkably helpful. Rick got his first job by approaching the provost of his undergraduate alma mater, Haverford College, with the argument that ecology should be offered and that he was the person to do it. It seemed like a huge and scary long shot, but it worked, and we have witnessed similar successes repeated in various employment situations.

There are many sources of information about the nuts and bolts of putting together a successful job application and preparing for your interviews. We will not cover this process here but recommend Chandler et al. (2007) for a focused and relevant guide to creating a strong research CV and statement of research interests. We will, however, share some tips about applying for teaching and nonacademic positions.

Teaching jobs

There are more jobs at teaching-focused institutions than at research universities. If you are getting your degree at a research university, you may well be getting the wrong idea of what will be expected of you when you go on the market for a teaching-focused job. In addition to demonstrating that you understand ecological research, you will need teaching experience. A recent quantitative analysis of successful candidates for tenure-track positions at teaching-intensive institutions revealed that prior experience as the instructor of record (not just as a TA) was virtually essential (Fox 2020).

Most graduate students receive little training that prepares them for teaching careers (Golde and Dore 2001, Austin and Miller 2020), so the idea of going from zero to being the instructor of record might sound overwhelming. Work your way up slowly. First, be a TA (teaching assistant), then offer to guest-teach one class session in the course you are TAing or in a course your major professor teaches. You might want to ask if the professor will share their notes to speed up the process of developing your lesson plan. When you're ready to teach your own course, you still may find it difficult to land one. Make opportunities by teaching summer-session courses at your institution or courses at local community colleges. Perhaps a professor at your old undergraduate institution is thinking of taking a sabbatical leave. (Of course, finding time to develop your teaching skills is important but will do you no good if you don't finish your research project, so don't let it get in the way of doing your research and writing up.)

Your job application will ask for a statement of teaching philosophy (SOTP). All of the cover letters and SOTPs from applicants for teaching jobs say how enjoyable, valuable, and rewarding teaching is, so this statement will not set you apart from the herd. There are several questions you will probably want to address in your SOTP. What do you want your students to learn or be able to do? How do you help them get there? How do you assess whether you've succeeded? Finally, how do you teach *all* of your students, not just the ones who will succeed no matter what you do? Give examples throughout. If you want new ideas to answer these questions, look at research on how to teach effectively. Search keywords like "active learning," "problem-based learning," or "flipped classroom" in *Web of Science* or *Google Scholar* to see what ecologists are doing to improve learning. Then practice those techniques yourself. Your statement of teaching philosophy will stand out as genuine and meaningful.

Regardless of which academic job you apply for, part of your application may include a statement of diversity, equity, and inclusion (DEI). DEIs have become popular for good reason; academia and society at large have long and ugly histories of systemic discrimination. These statements are usually written in first person. It is up to you whether or not to disclose anything about your own identity. For more material and to (continue to) build empathy, listen to speakers and read authors who explain their personal experiences with marginalization in academia. Read widely; the histories of misogyny, racism, ableism, homophobia, and other -isms vary. Spend time reading about universal design, stereotype threat, systemic racism, microaggressions, gender bias, and how American academic institutions have profited from unratified land treaties with

Native Americans and the forced labor of enslaved people. DEI statements are relatively new, so it's not always clear what questions you should be answering, but two to consider are: What are your values related to diversity, equity, and inclusion? What have you done and what will you do to make sure both colleagues and students of all backgrounds are included and supported?

Nonacademic jobs

As you go through school, it is easy to imagine a career in an academic setting since this is what you been exposed to. It is also what your mentors currently do and may be the only career they have ever known. A decade ago, roughly two-thirds of all jobs held by recent graduates with PhDs in ecology were academic (Hampton and Labou 2017), though it is surely a lower proportion for those without a PhD. But what about that other third of ecology jobs out there? How different are they? And how do you get them?

In nonacademic positions, success is measured in very different ways, depending on the position. For example, Ian works for the US Geological Survey, and the expectations for his work are very similar to someone working at a university. In contrast, we have colleagues in other agencies and organizations who do ecology but are judged by their progress in recovering an endangered species or by their ability to distill ecological research for legislators, thoughtfully run statistics for other people, manage restoration programs, or communicate with and organize communities concerning environmental issues. The day-to-day experiences of each of these people differ considerably from one another and from most academic ecologists. As a practical matter, it is not always clear what the expectations of nonacademic jobs are going to be prior to taking them. So, when considering and applying for these jobs, it is a good idea to ask a lot of questions about the work, expectations, culture of the organization, and so on, as best you can.

Most nonacademic jobs focus more on problem-solving than on concepts or creativity. In contrast, academic training in ecology focuses largely on creativity and concepts (identifying problems and linking those problems to broader concepts) and not on solving previously described, practical problems. This distinction may sound, well, academic, since problem solving demands concepts and creativity. Nonetheless, the tone and emphasis will likely be different from the one you experience in academia. If you want a job outside of academia, it may be worth your time to pick a research question that not only satisfies your degree requirements but also gives you practice solving a real-world problem (Schwartz et al. 2017). A past program officer at Ian's job always asked for an "elevator pitch," a succinct description that is compelling to non-scientists (legislators, ranchers, bureaucrats) who want the project to solve a real problem. Being able to frame your work in this way when applying for nonacademic jobs is probably a good idea.

If you want to work for a nonprofit or government agency or in the private sector, you will make yourself much more attractive if you gain some experience with that outfit before it's time to apply. Many grad students who get nonacademic jobs ask scientists who are working in the relevant job sector to serve as either formal or informal committee members. Making connections with potential employers while in school is valuable; internships or research collaborations are two excellent ways to do this.

Finding nonacademic jobs is not that different from finding academic ones. If you know a group that you'd like to work for, check their website for job postings. For US government jobs, this is USAjobs.gov; comparable sites exist for state agencies and larger nonprofits. Many nonacademic ecology jobs are posted on the Ecolog-L listserv alongside academic ones.

No matter what type of ecology career you are pursuing, finding a satisfying job can be a difficult process. Some people pay dues for years (multiple postdocs, visiting instructorships), while others seem to luck into good jobs quickly. Watching people over the years, we have seen that persistence pays off. People who keep improving their application packages seem to end up with jobs they enjoy.

Science is a far more social endeavor than we had imagined when we entered the field. Effective communication with people around you can seem difficult and not relevant to doing ecology, but it can be as important to your success as the science itself. Even though your main focus is figuring out how to do research effectively, it's well worth also investing time in developing the other skills and contacts you'll need to get the type of job you really want.

CHAPTER 8

Communicating
What You Find

Learning about nature is fun, but the field only advances when you communicate what you have learned. Research and communication are very different skills. From society's point of view, if you don't make other interested people aware of what you have learned, then you haven't learned anything.

Not all attempts to communicate are equally successful, and this aspect of ecology has an enormous effect on whether your findings and ideas will have an impact. In the sixth and final edition of *The Origin of Species*, Charles Darwin included "an historical sketch of the progress of opinion on the origin of species." He explained why his ideas really were different from those of numerous predecessors who, by 1889, wanted some of the credit and fame for the theory that Darwin had expounded. Most of the authors were easy to deal with; they had simply missed the main points of the theory of natural selection. However, one author was more troublesome for Darwin, and he wrote, "In 1831 Mr. Patrick Matthew published his work on 'Naval Timber and Arboriculture', in which he gives precisely the same view on the origin of species as that (presently to be alluded to) propounded by Mr. Wallace and myself in the 'Linnean Journal', and as that enlarged in the present volume. Unfortunately, the view was given by Mr. Matthew very briefly in scattered passages in an Appendix to a work on a different subject, so that it remained unnoticed. . . ." Matthew understood the principles and their significance, but he didn't effectively communicate what he had grasped. He had the same impact as if he had never had the ideas in the first place.

This problem is not limited to Victorian times. For example, MacArthur and Wilson (1963, 1967) revolutionized the field with their theory of island biogeography. Years before MacArthur and Wilson, Eugene Munroe had proposed the same equilibrium theory, along with empirical support for the species-area relationship for butterflies in the West Indies, and detailed models to explain it (Munroe 1948 [his thesis], 1953 [an obscure proceedings]). Unfortunately, Munroe did little to communicate what he had found, and the scientific community remained unaware of his insights (Brown and Lomolino 1989).

These examples illustrate that it matters how and where you publish. Some journals are more influential than others and reach a much wider audience. Send your manuscripts to the most-read journals that are a fit for your research. Both Matthew and Munroe are forgotten footnotes in the history of ecology because better communicators independently came to similar conclusions. How many Matthews and Munroes have there been, whose potentially revolutionary advances have never been repeated or communicated? Send your work to a journal where it will be seen by other interested ecologists. On the other hand though, if you get rejected, remember that's just part of the process (see below) and send it somewhere else. Getting it out there is better than letting it languish in a forgotten corner of your computer.

The process of putting together a manuscript, poster, talk, or grant proposal helps you figure out what you know, what you don't know, and how the various pieces fit together. Almost all seasoned ecologists will tell you that they often think that they have a pretty good grasp on a subject they are about to lecture or write about. However, once they sit down and look for the actual words they are going to use, they realize that they haven't thought the ideas through. The act of writing or speaking clarifies your thoughts and will probably be valuable for you, independent of the other values of communication.

Writing your work up isn't a magical process. A few specific techniques can help you produce a strong manuscript, talk, poster, or research proposal much more easily than if you go about it haphazardly.

It might not be intuitive, but it's best to tackle the different parts of your manuscript, poster, talk, or proposal completely out of order. Assuming you have results already, start with those. They're the cornerstone of the whole thing. (If you haven't started your field season, identify your expected results. You probably won't get them, but this step lets you think about what your current experimental design can and can't teach you.) Then decide on your take-home message, which is just one or two sentences and should follow directly from your results. After that, the order is less important. The abstract should come last. It becomes almost fun to write after you have the rest of the paper together.

Breaking writing down into steps

In addition to planning to work out of order, we also recommend that you see the writing process as several distinct steps. You might expect that you should sit down and write a polished version the first time around. This idea, called "perfect drafting," has the power to destroy your productivity and peace of mind. It's probably a slower way to write your manuscript than waiting for an infinite number of monkeys to do it. Instead, whether you are preparing a manuscript, talk, poster, or grant proposal, think of communicating as a process with different goals and audiences at each step. Including all four steps (below), if you don't already, should make the process less difficult. See a summary in box 5 and a set of recommendations below.

Step 1. Pre-writing

It's hard to work out your thoughts and think about what your audience needs at the same time. So, before you write for others, write for yourself. Begin by gathering your information and creating an outline.

By "gathering your information," we mean rounding up your statistical output and figures, the articles you plan to cite, and your field notes. If you take our advice from chapter 6, you will also have a write-up of your methods and results after each field season. A preliminary lit review for the introduction and conclusion gives you a sense for how

Box 5. *Steps in the writing process*

Step	Goal	Audience
Pre-writing	· Organize your information · Create an outline	yourself
Drafting	· Write a messy draft	yourself
Revising	· Fill holes in logic · Remove everything not part of single story · Add anything required for complete story · Reorganize as necessary	others
Editing	· Fix diction, grammar, and punctuation · Get language feedback from native speaker	others

your work fits into the bigger picture, and your preliminary results section makes it clear what you have nailed down conclusively and what parts of your argument are weak and need further testing. This write-up will give you a big head start on an outline for your journal manuscript, talk, poster, or even grant proposal.

Next, we recommend that you create a fairly detailed outline. It can be either formal with roman numerals (Rick's preference) or informal with bullets (Mikaela's preference). When we're organizing our work, we use a list when we aren't sure about how to order the various ideas. Then we give each of the ideas a number or color code that helps to group the ideas that are similar or related. Next, we decide which of these should go first and how to connect them to make a logical argument. If you think you don't like outlines but haven't actually used one to write a professional paper, we recommend you give them another try. They make you more efficient in the long run and help you write a more organized paper.

Step 2. Drafting

Once you have your building blocks, including your outline, writing a draft should be much easier. For the first draft, just get the ideas out of your head. Your goal for this round is to figure out what you know and want to communicate. This is the point at which some people try to "perfect-draft." If you catch yourself doing that, move on. For now, you want a messy draft, nothing more. You are not writing this for readers or conference-goers, just for yourself.

Step 3. Revising

After you have a full draft, let it sit to gain a little distance. Taking some time away will allow you to see your draft as an outsider might. Your goal now is to make your target audience want to read what you wrote or listen to what you say.

Sometimes Rick tries to imagine that his father is reading the manuscript. His father had no formal training and therefore didn't have the jargon and preconceptions of a trained ecologist. Rick asks: Would he follow what I am saying? Did I provide enough background? Did I connect the ideas logically? Your draft should (in most cases) contain only one story, and that story should be as complete as possible.

As you revise, bolster the story format. Maybe the idea of telling a story sounds like a technique for fiction, but humans have been listening to stories much longer than to research talks, so using similar principles can keep your audience engaged. Stories need a good hook—spend time making the opening as engaging as possible (discussed in the section on introductions below). Stories have an arc—build up your supporting information in a logical and compelling order. They have a climax—make your results feel exciting and explain how they answer your initial question. Finally, a great story often has some kind of moral or bigger picture payoff—in the case of your scientific study, highlight your take-home message. For detailed advice about writing scientific papers as stories, we recommend Schimel (2012).

This stage may catalyze large-scale changes and will probably require several iterations. You're done when you've told your story in the most logical and compelling way possible for your expected audience.

Step 4. Editing

When you are finished with your revisions, then and only then is it worth your time to make sentence-level edits. Do that now, with an eye toward not only diction, grammar, and punctuation rules, but also any small changes that will make it easier for your audience to slog through yet another article, talk, poster, or grant proposal. Your audience is overwhelmed with information from all sides. Do your best to make your research easy for them to digest and appreciate.

If your native language is not English, you probably already know that it is a good idea to get a native speaker to help you with this last step. (Native speakers: you were lucky enough to get tens of thousands of hours of free English tutoring in childhood. When you review someone else's manuscript, differentiate between the quality of their English and the quality of their research. Focus your suggestions on how to make sure that their research gets read.)

In the sections that follow, we will offer some more specific suggestions for communicating your research in the forms of (1) a journal article, (2) an oral presentation, (3) a poster, and (4) a grant proposal.

Journal articles

Journal articles are the bread and butter of biologists. Writing papers can seem daunting at first, but as you begin to recognize the formula for writing them, they will become easier to write. Journal articles serve the important functions of archiving what you have learned and making it available to share with the rest of the ecological community.

Most journal articles are expected to follow a standard format: abstract, introduction, methods, results, discussion, and conclusion. (Even articles in *Science* and *Nature* are written in this format, although it is less easy to spot.) Each journal has its quirks. Read the instructions for authors carefully before submitting. In the following sections we give you some information that will demystify the formula.

Title and abstract

The paper starts with a title and an abstract. The title tells what the paper is going to be about. Don Strong, a former editor-in-chief of Ecology, says it should present the main result rather than just including the keywords. For example, "Fire increases butterfly diversity in riparian and woodland habitats" is a more informative title than "Effects of fire on butterfly diversity in two habitats." Titles that sound clever but cryptic, such as "Diversity up in smoke," are a bad idea, as they convey neither keywords nor results.

The abstract provides a summary of the paper. As such, the abstract should include a sentence or two of rationale, the main results, and what your results mean in a broader context (in that order). The abstract must be concise and clear. Far more people will read your abstract than other parts of your paper. Even if they do read the entire paper, reviewers and critics of all kinds will make their decisions about the paper and your story based largely on the title and abstract.

Even though these come at the beginning of the paper, we recommend that you write them last, which is when you have the clearest sense of the main points and their significance.

Introduction

Your introduction should present your research question in general (conceptual) terms and explain why it will be interesting to your audience. You should immediately "hook" your reader and set the stage for your question. The most impactful papers tend to start by stating a general and important problem (understanding climate change, coevolution, parasitism vs predation, and so on) in their very first sentence. They then point out a gap in our understanding of that problem and how their research will fill it. At the broadest level, what issue is at stake in your research? What is missing from the current literature on that issue that your research addresses? If you convince readers that it's important or if you pique their curiosity, they are more likely to read on. Make sure that this hook is tailored to your audience (e.g., conservation ecologists, basic physiologists, theoreticians, etc.).

We find an introduction that poses a big-picture question to be much more effective than starting with a description of your organism or study system (although the latter is how many students think they are supposed to begin). So, for instance, don't tell us that you are interested in wooly bear caterpillars. Instead, as mentioned above, lead with the general question: What are the consequences of food choice? Then tell us how your study will provide an answer to a problem that hasn't already been solved. For example, explain that food choice may allow herbivores to survive parasitism, and that your study organism, wooly bears, will allow you to answer this question. Similarly, don't just assert that it is critical to know the rates of predation of lemmings. Instead, start with a brief review of the intriguing phenomenon of population cycles. Then explain how predation will help us understand the drivers of herbivore population cycles. In other words, make sure that you have explained why we as ecologists will want to know what you have found. Don't just tell us you will fill a data gap—explain why that data gap is interesting.

Reference other research that informs your big-picture question, such as prior work on the question that has been conducted in other systems. If you followed our advice from chapter 6 about reading the literature efficiently, you already have a few sentences for each of the papers you've read, summarizing the relevant findings. This is the time to take advantage of those notes.

Sometimes students aren't sure how to choose which articles to cite. It's not possible to cite every article related to your study; the literature is just too vast. You don't have to cite an article if it states something that is widely accepted or if that paper has not been influential to the field. You usually need to cite a paper only if you are talking about the results of the study or if you are borrowing a particularly interesting idea. When choosing which article to cite, you will almost always want to start with the most influential paper on the topic—often the first study to have the finding you are discussing in that sentence or paragraph. For example, if your study involves island biogeography, you should probably cite MacArthur and Wilson, but citing Munroe is optional (see story above). You may also want to cite a review or the most recent similar finding.

As a rule of thumb, don't cite more than three papers per statement; it's usually best to cite fewer. One exception is controversial statements; in those cases, you may want to include more citations.

Don't include references just to show that you are familiar with the literature. Paradoxically, that will weaken rather than strengthen your manuscript. Instead, cite those papers that motivate and explain your question and those that highlight the importance of what we don't yet know (such as a review or an idea paper).

When citing published work, describe the results and not what the authors did. For instance, say "Females were 30% bigger than males (Brown 2020)" rather than "Brown (2020) reported that females were 30% bigger than males."

Your introduction should contain an explicit statement of your question or hypothesis—tell your audience exactly what you are asking. We often find figures or cartoons that represent ecological hypotheses to be informative. These can be simple path diagrams that describe hypothesized causal schemes, or more complex figures that provide a visual representation of an ecological model. Many readers grasp concepts more readily when such figures are included alongside verbal or mathematical representations.

Explain that you will test the general question by looking at a specific example. Details about the natural history of your system may be useful here. You may also want to explain how research in your system specifically will contribute to answering the more general question.

We like to end our introduction by giving either a formal listing of the specific questions that we will answer (for a paper) or a brief answer to the question that we posed at the beginning (for a talk; see next section). (The questions don't have to be stated as null hypotheses, which have a bit more statistical precision but can be confusing to follow.)

Presenting the questions in such a prominent place lets the audience see where you are going to go in your methods and results. Also, because this formula is relatively standard, some readers will skip to the last paragraph of your introduction to decide whether the questions you ask are interesting enough to keep them reading.

Finally, some authors write the introduction in the present tense, describing the current state of knowledge (though the results of specific studies are generally in the past tense). The rest of your manuscript should be written in past tense since you are describing what you did, what you found when you did it, and how you interpreted those findings.

Methods

The methods section explains what you did to answer the questions that you posed in your intro. Make sure that the order and wording in your description of the study design aligns closely with your research questions.

Methods sections often begin with descriptions of the natural history of your system. (This can also go near the end of your introduction, right before the list of questions that you will answer.) Tell us enough, but only enough, natural history so that we can follow the important points of your experiments and interpretation.

Your methods must be described clearly enough that the work could be repeated by someone else. This should include a description of where, when, and how you applied your treatments and took your measurements. Make sure that the reader understands the motivation for each experimental procedure; instead of just launching into the

details, start the description of each experiment with something like, "To test the hypothesis that wooly bear caterpillars choose the most common host plants. . . ." At the end
of each methods subsection, include a brief description of
the statistical analyses, including the statistical package,
that you used.

Your methods section should be kept brief; provide
only enough information so that another person could re-
produce the study. Graduate students often assume that
they need to explain every detail of their projects, but this
is not true. Only include information that is directly relevant to the story that you want to tell. For example, perhaps
you kept detailed data each day on temperature, percent
sun, or precipitation because you thought that it might
help explain variation in crawdad feeding events. However,
you didn't find any relationship, and your story moved in
other directions. Your readers don't need to hear about
these details even though you may be tempted to show them
how thorough you have been. Information that isn't immediately relevant will only clutter up your presentation
and obscure your main story.

Results

The results section tells what you have found—your data
and statistical analyses. In general, you will not be able to
get away with making a statement if you don't have statistical analyses to support it. Results should be described in the
way that tells your story logically. Often students describe
their results chronologically, in the order in which the
experiments were conducted. But a topical organization

that answers your main questions in a logical progression is generally more effective. We like to organize our methods and results sections by questions or experiments. We use the same subheadings for each question in both the methods and results sections; the subheadings and subsections are repeated in order in both sections so that the reader can easily connect a method with the corresponding result.

Data should be presented in the text or summarized as a figure or a table. Text or figures are more effective at conveying most kinds of information than are tables. Text is the default choice; if a result can be conveyed effectively by describing what you found, do so. Figures are particularly good at conveying relationships between factors, although actual values are generally obscured. Tables show exact values but are not good at presenting relationships between factors.

Often the most informative way to present your results is to illustrate your data in a figure and then describe the result and effect size in the text. For example, if you are describing the effects of selenium on salmon fecundity, you might use a bar graph to show that female salmon produced approximately 60 eggs in low-selenium streams and 20 eggs in high-selenium streams. Then, report the percent change in the text: "High selenium levels dramatically reduced salmon egg production. Female salmon produced only about one-third as many eggs in high-selenium streams as in low-selenium streams" (followed by results from a statistical test comparing these means). Make your results, not your tables and figures, the star of the sentence.

The tables and figures merely illustrate what you are describing in the text. "Adults consumed 40% more than juveniles (table 1)" is preferable to "Table 1 shows the consumption rates of adults and juveniles."

Your results should be presented in biology-speak rather than statistics-speak, and you should highlight the biological results and not the statistical tests. For instance, tell us, "Drab individuals were twice the size of iridescent individuals (Student's t = x, df = y, $p=0.0z$)" rather than "The Student's t-test with y degrees of freedom showed a statistically significant effect at the 0.0z level of iridescence on body size." Also, always present the effect size, not just the level of statistical significance—"Drab individuals were twice as big as iridescent individuals" rather than "Drab individuals were significantly bigger." The effect size tells us about the biological relevance of the result, whereas the statistical significance tells us how likely it is that the result was due to chance.

Simple figures and tables are better than more complicated ones. Figures and tables should be described clearly with titles and captions, so that someone looking at them can make sense of the results without necessarily reading the whole paper. Your audience must be able to discern exactly what you measured. In figures, this is often achieved by clearly labeling the axes and including units. The fewer the number of treatments presented in one figure, the easier it is to grasp. Don't combine figures unless viewing all the information at one time adds meaning. If possible, identify the treatments within the figure rather than in the caption.

The most commonly used figures are bar graphs and then scatterplots. When you use bar graphs, it is easier for a reader to grasp the meaning if you use fewer bars. Under most circumstances, error bars should be presented with all bar graphs. These are important because they give the reader a sense of the noise around the signal. You will usually use standard errors to show the noise or precision around your estimate of the mean. Standard deviations are used only when you want to show the amount of variation *per se*. Occasionally, error bars make a figure so busy that the signal becomes unrecognizable, and only under these circumstances should error bars be omitted. Scatterplot figures are also commonly used by ecologists. If the model is found to be significant, the line that describes the best fitting model can be added to the scatterplot.

When you get to the point where your manuscript is accepted, the publisher may hound you about the readability of labels and lines you use in your figures. Do yourself a favor and anticipate this requirement. You may need to make your font larger and lines thicker than your software's defaults.

Tables should be used only when repetitive data are essential to tell your story. For many arguments, fewer data are more effective than more. Only include those variables that are relevant to your story. One common application for tables is to summarize statistical tests. For example, in an analysis of variance, the sums of squares, degrees of freedom, F ratio, and p-values all provide unique information. If this information is not required to make a convincing

case, then include just the F ratio, degrees of freedom, and
p-value in parentheses in the text.

As we mentioned in the section on methods, it may be
tempting to include all of the experiments and observations that you performed. Don't do it. Include only those
results that are connected logically—that tell one coherent story. Variables and effects that are not relevant to the
story should be omitted, or the audience will be distracted
from the main points. Many authors make the mistake of
trying to include all the data they have instead of thinking
about what pieces are needed to tell the best single story.
Also, don't use the results section as a core dump for your
field notebook. If you feel compelled to include data for
archival purposes, stick them in an online supplement
rather than bloating your results.

Discussion

In the discussion, you should explain what your work
means in a broader context. To do this, it is often a good
idea to restate concisely the most important result before
you interpret it (this can be your opening sentence). How
do you make sense of what you found? Do your results contribute to the resolution of the question that you posed in
the introduction? What evidence have other studies
brought to bear on the question? Then as they become relevant to your story, add in the other results of your study
and interpret them. Often the results of experiments will
suggest subsequent hypotheses, which can be integrated
into your discussion now. You may be able to generalize
from your results in conjunction with those of others. Do

any useful models or paradigms result from this work? You started your paper by including a hook that addressed a general question. Your discussion should return to that question and provide an answer in equally general (conceptual) terms. This doesn't mean extrapolating from your system to all others in the universe, but it does mean thinking about the results from your system as a model or example of a larger answer.

We often see discussion sections that revolve around what the authors expected to find rather than their actual results. Believe your results and interpret them as such. If you didn't find what you were expecting, remember that most discoveries are surprises. If you already know the answer, then the question isn't particularly interesting. Don't talk about how your results might have been different with a larger sample size, or if you had controlled for other factors, or if you did the work in a different place. Instead, interpret the results you actually got.

Throughout your paper you should tell a cohesive story. Don't wander from your central point. Rather, your writing should present a tightly reasoned argument that is evident from start to finish.

Conclusion

The function of the conclusion is to make sure that your main finding doesn't get lost but instead is absolutely clear to those people who will scroll straight to the conclusion or have gotten bogged down in the details. So, papers should end with conclusions (though they often don't, which confuses us). You'll see papers that end with a non-conclusion, like "this is a good system" or "more work should be done." Of course, more work should be done following every study. Notice how uninspiring this is to you as a reader. Instead, leave us with what you have learned and its consequences.

To do this, first, state once again the questions you posed at the start, followed by your main results. Then, state the consequences of your findings. Put them in the larger context. Explain that in a few sentences or up to a paragraph. Sometimes, by the time you've done this, you've already stated your take-home message. If you haven't, do so now. Don't have too many take-home messages, or your reader won't remember any of them. If your reader can only remember one thing from your manuscript, what should it be? Make sure you hit them over the head with that.

Box 6 provides a summary and checklist of our suggestions about writing journal articles.

The publication process

The publication process can be emotionally brutal and requires a thick skin. All ecologists get their manuscripts rejected. Ecologists who succeed at publishing the most also experience the greatest number of rejections (Cassey and Blackburn 2004). Even established professors experience a rejection rate of 22%. In a careful statistical analysis across multiple journals, papers that were rejected, revised, and submitted to a different journal were ultimately more often cited than those that were accepted by the first journal (Calcagno et al. 2012). The reasons are likely complex. For example, papers that contain novel ideas

Box 6. *Journal article checklist*

Title
☐ Does the title summarize the main result?

Abstract
☐ Does the abstract tell your story concisely?

Introduction
☐ Does the beginning of your introduction "hook" the reader?
☐ Do you open with the general (conceptual) topic your research addresses?
☐ Does your intro clearly identify the gap your research fills?
☐ Do you explain and justify your question(s) instead of just extolling the virtues of your study organism?
☐ Do you briefly summarize previous work that informs your question(s)?
☐ Do you end your introduction by clearly listing the question(s) your manuscript addresses?

Methods
☐ Do you briefly explain the relevant natural history of your organisms or study system?
☐ Do you describe your methods thoroughly enough that another ecologist could repeat your work but briefly enough that space-pressured journals won't send your manuscript back?
☐ Do you include a brief explanation of each statistical procedure you used?
☐ Do you include only the methods relevant to your overall story?

Box 6. Continued

Results

☐ Are your results presented in a logical order to help your reader follow your story (not in the order in which you did the experiments, if this is different)?

☐ Does the text inform your readers of your results as much as possible instead of simply referring them to your figures or tables?

☐ Do you present effect sizes for each of your results?

☐ Do you describe your results in biological rather than statistical terms?

☐ Do you present each of your results as an integral part of your overall story?

Figures and tables

☐ Do the titles and captions convey enough information that your reader can understand the figures and tables without looking at the text?

☐ Are your figures and tables as simple (decluttered) as possible?

☐ Do your figures show error bars, if appropriate?

☐ Do your scatterplots include a line that describes the best fit, if appropriate?

☐ Do you identify the treatments within the figure, if appropriate?

☐ Have you trimmed your figures and tables to the lowest number possible to tell your story clearly?

☐ Are your figures print-ready, according to the webpage of the target journal?

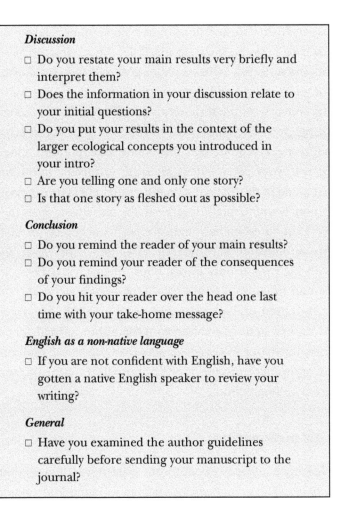

Discussion
- ☐ Do you restate your main results very briefly and interpret them?
- ☐ Does the information in your discussion relate to your initial questions?
- ☐ Do you put your results in the context of the larger ecological concepts you introduced in your intro?
- ☐ Are you telling one and only one story?
- ☐ Is that one story as fleshed out as possible?

Conclusion
- ☐ Do you remind the reader of your main results?
- ☐ Do you remind your reader of the consequences of your findings?
- ☐ Do you hit your reader over the head one last time with your take-home message?

English as a non-native language
- ☐ If you are not confident with English, have you gotten a native English speaker to review your writing?

General
- ☐ Have you examined the author guidelines carefully before sending your manuscript to the journal?

may be more skeptically reviewed because they are hard for some gatekeepers to assimilate.

Also, in general, peer review improves scientific papers, though it can be a painful process. Reviews from journals indicate how two or three readers perceived your paper. If they missed important points, other readers are likely to miss the same points. Take the comments of reviewers seriously. When a paper gets rejected, put it aside for a day or two, and then make changes that will address the concerns of the reviewers whenever possible. If you are being given the opportunity to resubmit your manuscript, address every point that the editor and reviewers raised both in your cover letter (referring to specific line numbers) and in the text of the manuscript. Letters from editors generally sound more negative to the uninitiated than they were intended to. Few manuscripts are accepted on their first submission. Having your submitted paper "rejected with the opportunity to resubmit" means it has been accepted (if you can address the reviewers' concerns).

As a grad student, you can learn a lot about the review process by getting involved yourself, either formally or informally. Offer to review papers for other grad students, for your major professor, or for journals if your professors are handling editors.

Oral presentations

Hearing a talk is a very different experience than reading a paper. The audience can't take in as much information when watching a talk as when reading a paper, so paring your story down to its essence is key. You'll find yourself giving talks in different circumstances—for example, your qualifying exam is a talk—but we'll focus here on live presentations for seminars or conferences.

Preparing content

Good talks require as much structure and planning as good papers. Don't imagine that you can just wing your talk; use some version of the four-step writing process above to prepare. Also, use the suggestions in the "Journal articles" section of this chapter as a guide to creating the components of your talk, but skip the methods section for now. (More on that later.)

When you give a talk, you are telling a story, but don't write it as a mystery. The best way to get your audience to remember your main result is to set it up right from the start. Organize your outline to tell the audience what your take-home is going to be, deliver the content, and then remind the audience what the take-home message is. Someone, either Aristotle, Dale Carnegie, or Winston Churchill, seems to have referred to this as "Tell them what you're going to tell them, tell them, and then tell them what you just told them." If you follow it, someone listening to your talk should be able to grasp your take-home message even if their thoughts drifted away at some point or they came into the room after the talk began.

Let your questions dictate the outline. Don't use the traditional methods, results, and discussion that you would in a paper. Instead, lead with the questions (or sub-questions) and nest the relevant methods, results, and discussion within them. For each question, explain in a sentence or two whether you did an experiment, a quantitative observation, or a model, and what factors were hypothesized to cause what outcome. Finally, describe your result in a sentence or two. What does this result mean? If it stimulated the next question, explain how. Repeat the sequence for each question, again making sure your audience can follow your logic.

Your talk must have a (1) single and (2) fully formed story to tell. If you've heard a talk that seemed like it was wandering, it may have violated one of these two ideals. Don't be tempted to include all the loose ends of your research—just your best single story. Try not to switch gears (that is, tell two stories) unless you can integrate them into one clear, cohesive story; your audience won't take much away from such a mixture. We have noticed that talks are far more likely to seem disjointed than manuscripts; for a talk, pay special attention to whether you have a single, coherent story.

If you don't yet have enough results for one coherent story, spend time thinking about how to organize what you have done thus far so that it seems like a single story. Your results will be much more memorable to your audience if you do this. How do you make it seem like a single story? First, analyze whether you are sharing the data in the most meaningful order. Second, find gaps in your story and see if you can fill them with results from other studies. Finally, smooth the transitions between the pieces. Ask yourself whether there is a clear path from your introduction and research question (or short series of interrelated questions) all the way through to the punch line. If not, keep organizing.

Preparing a talk is partially an exercise in subtraction. Documentation of the existing literature, methods, each step of your statistical analyses, etc., are crucial in a paper but stultifying in a talk. Minimize or eliminate them; include only information that is essential for you to tell your story (but be ready to talk about the rest if anyone asks at the end). A talk should not attempt to provide the listener with the ability to repeat the experiment. Keep each method as brief as you can, show it with a photo or illustration rather than words, and mention it immediately before the result it led to. If you need to include a citation, put it in a bottom corner of your slide.

Creating slides

Once you know the story you are going to tell, create your slides. Your slides should illustrate your story, not be your story. Plan to give your talk as if you didn't have slides. (In case there is a technical failure, you should always be able to give your talk without using slides at all.) When developing your slides, remember that your words and your slides are battling for your audience's attention. Your slides should support your words, not distract your audience.

Use the same four-step writing process we suggested above when you create your slides. After you've done the first two steps (pre-writing and drafting), think in terms of design principles for the last two (revising and editing). Ask yourself: What visuals do I need to help the audience understand and enjoy my talk?

Pictures are better than words or tables. In general, tables are much less effective in talks than in papers and should be used sparingly, if at all. Never show something that is too opaque for your audience to quickly spot its significance. Tables used in talks must be simple, with large, easily read characters.

Pick a non-provocative font (i.e., avoid Comic Sans) and stick to it. Your smallest font size should be 24. Use different font sizes to indicate the bigger picture and what to focus on. As a rule of thumb, put no more than 10–15 words on a slide. Use bullet points instead of a lot of words.

Aim for visual simplicity. Three dimensions and a cacophony of colors rarely help tell the story. In a recent study, viewers preferred and learned more from decluttered and focused figures (Ajani et al. 2021). Keep the graphics simple and the palette limited. Choose your palette with a colorblindness color generator (such as the one at david mathlogic.com/colorblind). If you must use a complicated figure or any type of table, ask yourself what you want your audience to focus on. Superimpose a colored circle or arrow on the slide to concentrate attention on that spot. Alternatively, you can use a bright color to highlight the result you are talking about and show everything else in dark gray.

Similarly, don't crowd your slides with too much information. Your audience will spend their time frantically trying to guess where to look. If possible, include only a single point per slide. You may have several figures or bullet points whose full meaning doesn't become clear until they are all shown together. In that case, to keep from overwhelming your audience, introduce the items to the slide one at a time (but don't make the new material fly in).

Spend enough time with each addition to let your audience process it.

This is especially true with complex information like theoretical or mathematical material. Be extra careful that your slides are clear and that the material is accessible. Use words rather than equations if you can. If your talk is mostly theory, you may have no choice but to include an equation or two. To help your audience focus on the relevant cluster of terms in your equation, add and then remove a circle or an arrow from your slide as you talk (see above). Explain the cluster in non-technical terms, and, if appropriate, add a simple graph showing what it means in a biological context.

Don't use slides with writing that you merely read aloud; but do make sure that the wording on the slides is complementary to what you are saying. Text on slides should be as abbreviated as possible—a few words or phrases, but not sentences. If you don't speak English often or are not confident with it, you might want to ignore this suggestion in favor of using a complete sentence on your question, main result, and take-home message slides.

Polishing

Your main goal so far has been to develop your content. Now it's time to start thinking from the perspective of your audience members. What do they need so that they can care about your research and remember your take-home message? Review the suggestions in "Step 3. Revising" above. Your goal is to turn your content into a story with a memorable take-home message.

Tailor your talk to your particular audience, thinking of their backgrounds and interests. Avoid jargon (abbreviations, specialized words, names of hypotheses or principles, measures, techniques) that is not familiar to almost everyone in your audience. Once people tune out or get lost, it is hard to get them back. You cannot expect people to grasp equations or complicated theory that they have not come to grips with previously. When readers come upon new or complicated *written* material, they can read and reread it until they understand it. There is little chance for this to happen in a talk, so don't lose your audience by speeding through this kind of material without a clear explanation.

Don't force your audience members to piece together the relationships between sentences or slides on their own. Instead, use signposts that help them see the structure of the talk and how each piece fits into the bigger picture. When you're reading a paper, you rely on the subheadings and paragraph indentations to recognize transitions. The audience at your talk needs similar large- and small-scale signposts. Oral repetition alone won't help most people remember complex messages (Kassim et al. 2018), so make sure you back up your signposts with key phrases on your slides.

Large-scale signposts prepare your audience for the big-picture journey, which we recommend you provide in the form of a bulleted outline. It helps people follow your train of thought if they can see where you are going throughout the presentation. Don't write a superficial outline ("Introduction," "Methods," and so on). Instead, give your audience a heads-up about how your story will develop (that is, use your research questions, results, and take-home message to structure your outline slide). Rick likes to write his outline on the board if one is available. Then he refers to the outline at various times during the talk. Mikaela uses a more normal approach. She puts up an outline slide early in her talk and shows it at relevant intervals throughout, using bold or red lettering to remind the audience what she just discussed and how it relates to what is coming next.

Small-scale signposts help your audience understand how one idea links to another. With each slide, ask yourself what the audience needs to know to place that information in the larger context of your talk. For example, if you tested three explanatory mechanisms for a pattern, say so up front. Then before you present each one, say, "The [first, second, third] mechanism I tested was. . . ." If the results you just presented led you to ask a new question, don't just jump to the next slide with an offhand, "And then I did the next experiment." Instead, explain the connection. Also, plan the signposts necessary to present your data. Remember that your audience members are seeing your data for the first time. Put labels on the x- and y-axes of every figure, and plan how to explain what is displayed on each figure you present.

Practicing

Practice your talk out loud before you give it. The more you practice it (especially in front of a real audience), the better it will be and the more confident you will feel when it's time to give it. If an audience is not available, saying the actual words out loud is much better than thinking them. Mikaela used to think it was too embarrassing to practice a talk out loud with her housemates overhearing her. Then she learned it was more embarrassing to give a poorly practiced talk. Recording a practice (or actual) presentation will help you to learn your talk quickly and will also improve it if you use the recording to make changes. If Rick has to give a talk that he really cares about or if he is strapped for time, he records a practice and then listens to it several times (or at least has the recording playing) while he's doing other things. It is amazing how much this helps. If you're really brave, videotape your talk and review it to learn what aspects need improvement, including what distracting movements you may want to avoid.

If you are new to presenting, our advice is to slow down. When you first begin giving talks, your nervousness is likely to cause you to speak faster than you practiced. (Often even seasoned speakers go too fast.) A good remedy is to pause after making an important point. The pause provides emphasis and gives listeners a chance to absorb the last message. Stopping to take a deep, slow breath also helps you remember the big picture and to pace your talk.

The more you involve your audience, the more successful you will be at holding their attention and having them remember what you say. Don't read your talk; a reading voice is far more difficult to comprehend than a conversational voice. Practice your talk until you can give it without notes. Though, if you are worried about forgetting what you want to say, you can create an outline on paper or in the notes section of your slides as a backup. If there are specific essential concepts or points that you are prone to forget, link them to a particular slide. When you get to that slide (usually a picture), it is your cue to remember to provide a particular piece of your story.

As a rule of thumb, use about one slide (or, if the slide is animated, one piece of information) per minute of talk. Many speakers make the mistake of including far too many slides and needing to rush at the end or annoy the audience by going overtime. To avoid rushing, time yourself when you practice. If you are still going over time, revisit the step of paring down to the essence of your story. This can be hard or flat-out aggravating if you are giving a short talk, but your audience will remember your take-home message more easily than if you rush.

Spend extra time preparing your introduction and conclusion. These are the only parts that some of the audience will hear. Everyone is most alert at the start. Get them excited about your study by placing it in a broader context or by explaining how it contributes to answering a more general ecological question. In other words, explain why what you did should be interesting to someone who isn't necessarily familiar with your specific system. Similarly, if they haven't followed all of the talk, the conclusion should hit them over the head with the take-home message.

Once you feel comfortable with your talk, plan ahead for the Q&A (question-and-answer period). It's hard to deal with complex questions and comments on your feet, so anticipate what you might be asked. Practice your responses out loud. If you have data relevant to a question you think you might be asked, keep it handy on a slide behind your final presentation slide, then pull it up if someone asks you that question.

PRACTICE TIPS FOR SECOND-LANGUAGE SPEAKERS
(AND THEIR COLLEAGUES)

If you're not a native speaker of the language you are using, giving a talk can be even more intimidating. Our friend Toshiko Murata, a law professor at KUAS in Kyoto, Japan, has some humorous but sincere advice on this. She says the best way to get your audience to understand your English is to tell them something they really want to know. That will get them to listen up. (Tailoring your talk to what your audience will care about is important advice for native and non-native speakers alike.) So, as mentioned above, justify your talk well, make your results really clear, and craft a straightforward take-home message. The goal is to increase your comfort and confidence so that you can enjoy the experience.

Sure, this all sounds great, but ultimately you will be standing in front of an audience that's expecting English. Then, any practicing you've done out loud beforehand will increase your enjoyment. If at all possible, get ready to say your talk conversationally instead of reading it. If you will need to read your talk, practice making it sound as conversational as possible. Also, rehearse key terms until you have made them as clear as you can. If you suspect your accent is still hard to follow, you can put those words into your slides and then point when you say them. As you give your talk, it can be tempting to speak quietly or very quickly, but resist the urge. Your audience will enjoy your talk more if you speak loudly and enunciate.

If you don't have time to practice every single word of a talk that is not in your first language, pick out a few of the most important sentences in your presentation (the statement of your research question, your main result, your take-home message) and ask a native speaker whether they are easily understood and what you could do to make them clearer. Remember that you want their honest and patient feedback because you deserve to get your research out there as much as anyone else does. (If you are a native speaker, volunteer to do this for your lab mates and collaborators. We work with a lot of international colleagues and have found that it is consistently rewarding to hear the results of someone working half a world away from us.)

Even native English speakers can find the Q&A portion of talks intimidating (see the discussion above), so if you don't use a lot of English at your university, you've got another burden on your shoulders—responding to a questioner who may mumble, speak too quickly, or use jargon. You can say, "Could you repeat that?" or "Could you speak up a little?" All three of us are native English speakers, so Mikaela checked in with her Japanese colleagues for advice on Q&As. They recommend doing mock Q&A sessions with friends and colleagues as often as possible. If you find that you aren't happy with the answers you give during these practice sessions, use the same procedure we recommended above: practice your answers out loud at home until they become easy to say. (If you are a native English speaker and have a question after a non-native speaker's talk, respect the fact that they are doing a whole second job in order to answer you.)

Game time

It is a biological fact that dark rooms and dark slides put people to sleep; use white as your slide background color, and keep the lights on when you present. A bright room is much more important than photographs that show up well on the screen. Which would you prefer, having some large fraction of the audience dozing while really sharp pictures go up, or having the audience alert and attentive at the expense of some photographic quality?

Look at the people in the audience; eye contact will help to involve them. You'll probably have a strong urge to face the screen and talk to your slides (we do); instead, talk to your audience. If you need to look at your slides, point to the relevant part on the screen and then turn back to your audience before you explain it. Better yet, glance at your computer screen rather than turning your back on the audience at all. A physical pointer, your arm, or a meter stick are better than a laser pointer; if you must use a laser pointer, don't wiggle it about, circle the object repeatedly, or let it wander aimlessly across the screen. Ditto for your cursor if you are presenting by video.

Stay close to your audience. This lets you relate to them more effectively. If the podium is too far away, move it closer or don't use it. Come out from behind it and address the audience directly. Walk around a bit in the space you are given as well. It is amazing how much the simple act of walking from one side of the screen across to the other helps keep your audience alert and focused. Speak to the audience members in the corner farthest from where you are standing—it will help you remember to project your attention throughout the room instead of just to the first few rows. Figure out where your visual dead zones are in the room (usually in your peripheral vision) and look toward them as often as you look at the center of the room.

A talk from a person who is slightly nervous is often better than a talk from someone who lacks sufficient adrenaline. However, excessive nerves can make a talk difficult to follow, so be gentle with yourself. Remember that the more times you give talks, the easier they become. (We know this offers little consolation right now.) A mistake many beginning speakers make is to be self-deprecating and apologetic. Replace this with enthusiasm; your feelings become contagious.

One of the best parts of a talk is the Q&A session at the end. We already know what we have to say, but we're excited to hear the spins that other people will place on our results. Often new and exciting ideas come up in the questions after a talk. We sometimes ask a friend to take notes during the questions so that we don't have to remember all the suggestions. We like to leave a lot of time for this part (10–15 minutes for an hour-long talk and 2–5 minutes for a 12- to 15-minute talk).

If the room is large or the questioner soft-spoken, repeat the question for the audience before answering it. Make sure you understand a question before responding to it. It is fine to paraphrase the question and ask the questioner if you have it right. If you don't know the answer, you can say so (and it's better to say you don't know the answer than to make one up). You can also say that the point raised is an interesting one and you will think more about it or design an experiment in the future to test it. You might ask whether the questioner can think of a way to test the notion that they are bringing up.

Sadly, we have sometimes encountered questioners from privileged demographics who try to take over because they are taller, whiter, older, maler, or whatever than the speaker. (We recognize that not everyone in these demographics is necessarily a seminar bully.) Remember that the floor is ultimately yours and that you can cut them off. If an audience member is aggressive and won't give up the floor, you can say that you would be happy to talk more after the seminar is over and then shift your attention to the next questioner. By the way, remember not to be that person when you are in the audience.

Finally, if the talk you are giving is your oral exam, try hard not to be defensive when your committee members question you. Rick has sat in on countless oral exams, and he has noticed that students who fail them are generally not the ones who are least prepared but rather the ones who become defensive and argue with their examiners.

Box 7 presents a summary and checklist of our suggestions about talks.

Posters

Posters have become the most common medium at some meetings. Remember that people at meetings are burnt out.

Box 7. *Oral presentation checklist*

See also box 6 for a general checklist about communicating in ecology.

Introduction

☐ Do you structure your introduction around a general question that your take-home message will address?

☐ Do you eliminate most of the citations and other details you would include in a manuscript to help focus your audience's attention?

☐ Do you end your introduction by showing a slide that clearly indicates the question(s) you will address?

Methods, Results, and Discussion

☐ Do you lead with a question and integrate your methods, results, and discussion to answer it?

☐ Have you ruthlessly minimized your methods?

☐ Do you explain how the results of your first experiment generated your next question and experiment so that your audience understands the relationship between the parts as your story develops?

Conclusion

☐ Do you hit your audience over the head with one clear take-home message?

Figures and tables

☐ Do you show your results in pictures and figures instead of just describing them?

☐ For your figures, have you practiced indicating to your audience what the x- and y-axes are?

Box 7. Continued

☐ Do you minimize or eliminate the use of tables since they are hard to grasp during a talk?

☐ Do you have a plan for concentrating audience attention (using circles, arrows, colors, or a steady laser pointer) on each of your results?

Preparing your talk

☐ Have you practiced your talk (especially the introduction and conclusion) until you are absolutely comfortable with the information in it?

☐ Have you created roughly one slide for each minute of your talk?

☐ Have you timed yourself to make sure your talk does not go overtime?

☐ Are you prepared to give your talk without any slides at all in case of a technical problem?

General structure and presentation

☐ Are you able to give your talk (from the slides or an outline) without reading it?

☐ Have you practiced making eye contact with your audience (instead of with your slides)?

☐ Have you practiced moving about the room enough to keep the audience engaged?

☐ Have you carefully examined your talk for jargon?

☐ If your talk includes an equation, have you planned how you will make it readily accessible to your audience?

☐ Does your talk include signposts so that your audience can follow the structure you have created?

☐ Do you use a large font size (24 point or larger) and include very few words (10–15 maximum at a time per slide)?

☐ Do you use a colorblind-friendly palette for figures where color tells your story?

☐ Have you created a coherent structure so that people can follow your entire story?

☐ Have you removed all information that is not required to tell your story?

Do you enjoy reading a lot of fine print when you are viewing posters? We don't. Instead, we want the take-home message in a simple, readily available package. Most posters suffer from being prepared like manuscripts. Their structure should be much more like talks than like papers, and their word count should make them the equivalent of illustrated abstracts.

Keep your poster simple. Everyone who walks by your poster should immediately know your question and the answer. Those people who are interested can ask you to explain in more detail. If you'd like, you can generate a QR code or bring handouts or copies of your journal article for them to take away. Doing so can remind you to funnel the details there and keep them off your poster, where they don't belong. It is often helpful to include a photo of yourself and co-authors on your poster so that interested people can find you during the meeting. Include your contact information as well.

Since the poster sessions of most conferences are very large, you will need to compete for an audience. The title of your poster must be compelling, since it is the only thing many people will read. They won't usually be grabbed by a list of the characters (your study organisms or system) or the question. Instead, get their attention by sharing the main message. For example, fewer people will be drawn to a poster entitled "Interactions between student protesters and campus authorities" than one entitled "Chancellor defends pepper spraying of peacefully protesting students."

Your introduction should be no more than a few sentences stating the conventional wisdom or explaining the justification for your question. Next, present your results as pictures (figures and photographs) that tell your story. Move logically from one result to the next, making sure not to include more information than your viewer can easily and quickly digest. As with a talk, your methods can be represented with a photo or diagram, but leave off the details of your experimental design, sample sizes, and so on. After each result, you can include one sentence of "discussion" that makes each result more general or relates it to your big question.

At the end of your poster, emphasize your take-home message. You might want to accomplish that by including a sentence or two that answers the specific question that you posed at the start, using the same wording to keep it crystal clear. Alternatively, or in addition, you may want to add a sentence or two (but no more) explaining the significance of your results and how they fit into the big picture. If you label this part "Conclusion," it will probably be the section of your poster that gets the most attention after the title and the figure showing your results.

After you have worked out the pithy content of your poster, spend some time figuring out how to present it so that exhausted conference attendees can still take it in quickly. Put the title in an extra-large font so people can spot it at a distance. Lean heavily on (relevant) images to illustrate your points and help you cut down on text. It's best to use a white background, both because colored inks are expensive and because colored backgrounds annoy many readers. Lots of white space and a limited, colorblind-friendly palette can help focus the viewer's gaze and are easier on the eyes. Draw attention to your headings and main points through careful use of font sizes and colors. Once you like your content and design, line up your margins meticulously. Printing a poster is expensive; make sure to get feedback before you do so.

An advantage of presenting a poster is that you can walk interested people through your story instead of asking them to read the thing. In addition, they have the opportunity to ask you questions about points they don't understand or suggest other experiments and directions. As such, it seems like a good use of your time at the conference to hang out with your poster and interact with viewers as much as you can. Practice explaining the content of your poster clearly and concisely so that you are ready when someone stops to ask about it.

The poster that we have described contains less than one tenth the number of words of most posters at ecology meetings. It tells only a single story and does this using only "headlines." It contains no or few references and methodological details. It has figures and photographs but rarely

Box 8. *Poster checklist*

See also box 6 for a general checklist about communicating in ecology.

Title

☐ Is the title in a very large font to be seen at a distance?

Introduction

☐ Do you limit the introduction of your question(s) to a few sentences?
☐ Do you clearly present the question(s) your poster will answer?

Methods

☐ Is your methods section extremely brief?
☐ Are your methods shown in photos or illustrations rather than text?

Results and Discussion

☐ Are your results presented mainly as graphics (bar charts, scatterplots, etc.)?
☐ Do you show the differences in treatments with photographs where possible?
☐ Do you briefly explain the significance of each result?
☐ Do you present each of your results in terms of your overall story?

Conclusion

☐ Do you include a sentence or two briefly answering the question(s) that you posed at the start?

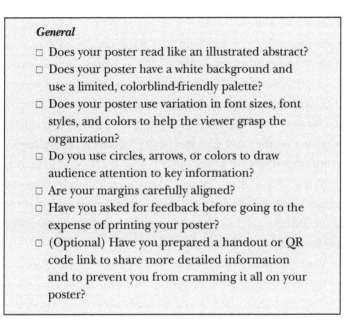

General

☐ Does your poster read like an illustrated abstract?

☐ Does your poster have a white background and use a limited, colorblind-friendly palette?

☐ Does your poster use variation in font sizes, font styles, and colors to help the viewer grasp the organization?

☐ Do you use circles, arrows, or colors to draw audience attention to key information?

☐ Are your margins carefully aligned?

☐ Have you asked for feedback before going to the expense of printing your poster?

☐ (Optional) Have you prepared a handout or QR code link to share more detailed information and to prevent you from cramming it all on your poster?

tables. It doesn't include a detailed description of your statistical methods or analyses, although statistical results can be useful if presented concisely. It relies on graphic design principles to help the viewer focus. It is much more effective at conveying information than the poster that is essentially a manuscript pasted on a board.

Box 8 presents a summary and checklist of our suggestions about posters.

Grant and research proposals

The purpose of grant and research proposals is to sell your plans about work you want to do. You want your committee to agree to give you a degree if you fulfill the objectives in your research proposal, and you want people to give you money in response to your grant proposal. In addition, your proposal provides two less obvious functions: it forces you to develop a research plan, and it forces other people to consider your ideas more carefully than they might otherwise so that they can give you better feedback. Nonetheless, grant and research proposals involve more salesmanship than research talks or papers. Therefore, a slightly different emphasis is required. As you prepare a proposal, focus on three things: (1) novelty and justification, (2) clarity, and (3) feasibility.

Your proposal outlines what you want to do. First, it must be exciting and original, and you must convince the reader that your work will advance your subdiscipline or the way in which people apply science to solve problems. Obviously, not every proposal is going to change the way all scientists think, but those people who work on your question or on your system should be influenced by your research. If it is not clear to you how this will happen, think hard about how to justify your work in these terms. Emphasize this justification throughout your proposal. If justifying your proposal sounds too vague, think about answering questions such as: What makes your proposed work significant? What is the value of the work? How might other people use your results? How will other people inside and outside of your field view the contribution of your work? The biggest mistake students make when writing proposals is not including enough justification.

Second, your proposal must be simple and clear, even more so than a scientific paper. Reviewers often get many proposals to read at one time, and all reviewers have better things to do than read them. Unlike scientific papers, these proposals may not be about subjects that the reviewers are already interested in or knowledgeable about. From the reviewer's first glance at your proposal, you have only a few seconds to convince them to pay attention and read further. Then you have only a few minutes to convince the reviewer that your proposal is worthy of funding from a budget that can in many cases fund fewer than 10% of the proposals in the stack. If your writing is not clear and concise, the reviewer may not do the work required to figure out what you are trying to say. The proposal must be convincing to both the meticulous reader and the rapid skimmer. A well-known (unnamed) colleague who serves on many NSF panels calls this the "two glasses of wine problem." He does all of his reviewing at the end of the day after two glasses of wine at dinner. Successful proposals must be clear enough to make sense to him under those conditions.

Third, you must convince readers that your proposal is feasible. Nobody is going to give you money or assurances of a degree unless they are convinced that you can complete the work you propose and that your work will answer the interesting questions that you have posed. There is an inherent contradiction in this process since your grant must appear both feasible and novel. You must convince people simultaneously that your ideas are important and ground-breaking, and that the work you propose can be accomplished. A great way to convince people that you can pull off your study is to use methods that have been used before and include citations for those methods. It is even better to be able to say that the methods are old hat for you or people in your lab. An excellent way to show that your plan is feasible is to present preliminary data. For a research proposal, this often involves first doing a year of fieldwork that addresses the main question. This heavy emphasis on preliminary results frequently means that researchers propose work that they have largely completed, and use the money to generate the next set of preliminary data.

Organizing your proposal

The organization of your proposal differs slightly from that used for talks and papers and is generally less rigid in its format. Different universities and granting agencies require different sections, and it is important to know these requirements and to fulfill them. Below we will consider the form and content of a proposal that would apply generally for many graduate groups and funding agencies. Many proposals include (1) an abstract or project summary, (2) an introduction, (3) explicit objectives, (4) experiments (or quantitative observations), justification, and interpretation that correspond to each objective, and (5) a budget. Other sections that are often helpful include separate discussions of the significance of the study, the potential pitfalls associated with your plan and your solutions to those pitfalls, a timetable for completion of each objective, and a justification of the budget. You can get more detailed advice about preparing proposals in Friedland et al. (2018).

The abstract or project summary shares many similarities with those discussed earlier for scientific papers. It comes first, though we write it last. It needs to be crystal clear and capture the excitement and rigor of the proposal. In it, describe the big-picture problem that you are addressing. Next, emphasize a justification for your work and explain its significance. Describe what you expect to find and explain why your results will be influential in your field. The project summary generally presents fewer results than abstracts for papers but can contain a few sentences about your approach.

The introduction to the proposal must get the reader excited about your questions. Justify why this work is important. This is difficult. Even after we've given such advice to grad students, more often than not their proposals could use more justification. How does your work relate to the big questions in ecology and why should we care? Frame your work in terms of the questions that you will address rather than systems that you will use. This advice holds even if you chose your project because you were interested in the system. In fact, it holds especially if you chose your project because of your system. You shouldn't write "This question is transformational" in your proposal. Instead, explain why it is transformational. For example, if you are studying tick population dynamics, say, "Tick-borne diseases cause x-number of deaths per year. To better control tick-borne disease outbreaks, we must first understand the population dynamics of ticks," and so on.

Your introduction should explain what has been done to date to inspire your question. It is often best to first present the general question and then describe how your specific research on a particular system will address that broader question. Start general and get specific. Here you can also tell us about the natural history of your system, but only if this information is immediately useful in understanding how you will answer your question.

Next, present your objectives. These can be long term (more than you can accomplish now to address a big-picture question) and short term (the actual goals of the work in this proposal). The objectives should be presented explicitly, and numbered. With each objective you should include a justification, and the hypothesis that will be tested. We will differentiate between objectives, justification, and hypothesis using the covid vaccine as an example. If the pharmaceutical companies Moderna and Pfizer submitted governmental funding proposals, their first *objective* would be to develop a vaccine that would provide resistance to the novel coronavirus. They would *justify* this objective by explaining that covid was highly contagious to a naïve human population and could result in millions of cases of mortality and morbidity that could be prevented with a vaccine. To fulfill this objective, the companies would test the *hypothesis* that messenger RNA could be used as a platform for effective protection. Their second *objective* would be to develop an mRNA vaccine. This objective might need a stronger justification—why not just use traditional vaccine technology? Their second *justification* would explain that mRNA vaccines would be safer and quicker to scale up for large populations. Finally, their second *hypothesis* would be that mRNA vaccines would cause fewer complications for immunocompromised individuals and could be quickly mass produced.

Each objective should be addressed by specific experiments or observations. It is often useful to number these exactly as you have numbered the objectives. A rationale and experimental design should be presented for each experiment or observation. Explain what each treatment will tell you and how it will be carried out, including sample sizes and controls. Demonstrate that you can accomplish each procedure. Finally, describe how you plan to analyze the data you collect.

Tell us how your results will be interpreted: "If experiment 1 gives this result, I will conclude the following." Interpretation of the results may or may not be its own section in the proposal. Remember that hypotheses in ecology must be testable, though they are not necessarily falsifiable or mutually exclusive. At this point you might want to include another section or paragraph entitled "significance" if the importance of your work has not been extensively discussed and stressed.

We like to include a small section detailing potential pitfalls. This section provides damage control and troubleshooting. Anticipate questions that the reviewers are likely to have and address them here. Try to explain how you will turn apparent misfortune into a situation in which you learn a lot. Describe here how you will interpret different outcomes from your study. The best projects are those that give interesting results no matter what the outcome. Try to design your study so that you are not dependent on getting

Box 9. *Grant proposal checklist*

See also box 6 for a general checklist about communicating in ecology.

General

- ☐ Have you explained to your readers why your proposal is novel?
- ☐ Have you explained why your work will be valuable to a larger ecological community?
- ☐ Is your proposal simple and clear, easy enough for an exhausted non-specialist to understand at the end of a long day?
- ☐ Is your proposal feasible, and have you explained this in a way that will be convincing?
- ☐ When possible, have you proposed established techniques and presented preliminary results?

Project summary/Abstract

- ☐ Does your project summary capture the excitement of your proposed research?

Introduction

- ☐ Do you take extreme pains to justify your proposed work?

Objectives

- ☐ Do you state each of your objectives explicitly?
- ☐ Do you justify each of your objectives?
- ☐ Do you articulate hypotheses that address each of your objectives?
- ☐ Have you designed and described experiments that address your hypotheses and objectives?

Interpretation, Significance, and Budget

☐ Do you describe how you will analyze your findings and evaluate each hypothesis?

☐ Do you highlight the significance of your potential findings?

☐ Do you include a budget justification, if appropriate?

☐ Are you truly excited to do the work if it gets approved?

☐ Are you *sure* you've provided enough justification?

one particular result to have something interesting to say. If you have designed a research program that will let you gain new and useful perspectives about nature no matter what the outcome, make certain that you stress this feature.

We also like to include a timetable for our objectives and experiments. This helps establish that we have thought about how and when we will get everything done. A timetable helps make the work appear feasible and is useful to refer to when doing the work.

Box 9 summarizes our suggestions about grants and research proposals.

Four things about the granting process should be kept in mind:

1. Grants are competitive, and it often takes several attempts before a grant gets funded. Don't get discouraged.

2. At the same time, take the comments to heart. We find it helps us get our emotions under control if we put the negative comments down for a few days before trying to deal with them. It can also be frustrating to get comments that seem to miss the point. If a reviewer missed our point, it indicates that we need to rewrite it (by clarifying the logic and details or improving the organization) so that even a sleepy reader with two glasses of wine in their stomach can follow. Almost invariably, the comments will contain extremely useful suggestions as well as these few misconceptions. If you are resubmitting a proposal, make sure to address all of the comments that you received from reviewers.

3. The granting process is very conservative (e.g., Boudreau et al. 2016). You can only get funding for ideas that everyone is already comfortable with. This hinders innovation. Over and over again, we have seen researchers change their priorities and pursue projects they aren't excited about just to chase a few dollars. As much as possible, don't let the granting process dictate your questions.

4. You may have no choice but to work within the boundaries of the granting process. You may be passionate about a question that cannot be answered cheaply, or you may be required to get grants to continue at your institution. Some agencies allow you to pursue additional questions with the funding they've given you for your main project. Also, smaller grants are often less competitive

than larger ones. Ask your major professor which
small grants may be available to you.

Ian notes that the granting process has become the
product rather than the means to an end. He suggests that
if you break out the champagne when you get funded but
not when you get a truly exciting result, something is amiss.

Our advice is to follow your own intuition and passion if
you can. Proposals are approved by a committee of scientists. Why give up something as personally important to
you as your research direction to an anonymous committee? Would you let a committee of scientific peers approve
your choice of a romantic partner over the next three to
five years? On a strictly practical (but anecdotal) level, we
have noticed that students who pursue funding-driven
rather than curiosity-driven questions seem less motivated
to battle through the many difficulties of grad school. If
you have the option of pursuing questions that intrigue
you personally, you may be better positioned to finish your
degree in a time frame you are happy with. If you are dependent on funding to remain in the field, try to create
some aspect of your project that really excites you.

Hard work often determines productivity, and productivity often determines success. Pick the questions that are
most exciting to you whether you get funding or not and
you are more likely to work hard enough to complete a
successful research project.

CHAPTER 9

Conclusions

There is a card game called Mao that is popular on several university campuses. One of the rules of Mao is that players cannot ask or explain the rules. New players joining the game must deduce the rules by observation and trial and error. A player who fails to follow a rule is given a penalty. Doing field biology can be a lot like playing Mao. The rules of field biology, and academia more generally, often go unstated. In this handbook we have attempted to make the unstated basic rules of the game explicit. You may or may not wish to follow the rules. But you might as well know what they are because you will face the consequences for choosing not to follow them. Unfortunately, life and ecology are both complicated, and there are *also* long-term consequences for following the rules too closely. Here we highlight some of the rules of our game as well as some of the potential costs of following them too assiduously.

Rule 1. Manipulative experiments are a powerful and highly respected technique to establish cause-and-effect relationships in ecology. Manipulative experiments lend credibility to your study.

> *Cost of Rule 1.* Experiments are only as good as the intuition that went into the hypotheses being tested. Make

sure you find time to know your organisms or else your experiments won't teach you much. In other words, make time for observations and natural history.

Rule 2. Inferential statistics can provide strong evidence when evaluating clear hypotheses.

> *Cost of Rule 2.* Ecology is not a science of truly falsifiable hypotheses and universal laws. Generate alternative hypotheses and evaluate the relative importance of each one. Recognize the arbitrary nature of inferential statistics.

Rule 3. A large sample size of randomly assigned, independent, controlled replicates increases the statistical power of your experiments.

> *Cost of Rule 3.* Replication comes at the expense of spatial and temporal scale and therefore of realism. Recognize that ecological processes are often scale dependent, so, whenever possible, conduct your study at multiple scales.

Rule 4. Observations and experiments should be planned in detail ahead of the field season. Your methods should be tightly aligned with your research question.

> *Cost of Rule 4.* Ecological systems rarely act the way we expect. Don't get trapped insisting on your initial question. Be opportunistic and pay attention to the directions that your system is trying to send you.

Rule 5. Scientists are expected to write proposals and apply for funding.

Cost of Rule 5. Grants are competitive, and the funding process is conservative. Unless you like administration, don't let writing proposals replace fieldwork for you. As much as possible, do the projects that are the most exciting to you even if they don't get funded.

Rule 6. The currency for researchers (including grad students) is publications. If you're a grad student who still holds an undergrad mentality that grades and classes are useful currencies, realize that the rules have changed.

Cost of Rule 6. Just as grades never did perfectly mirror what you learned from classes, neither does a long list of grants and publications perfectly mirror learning about nature, advancing the field, and contributing to society.

Unfortunately, the rules reward short-term goals that may not be consistent with your longer-term goals. The good news is that most of us got into this business because we like being outdoors and learning about nature. You can and should make your job reflect your interests. You are likely to have more control of this as your career advances. While you play the game, keep your eye on the big prize: your own personal and professional priorities. This is your life! You will be more successful if you're enjoying it.

Acknowledgments

We have gathered the advice of our teachers, role models, and colleagues together in this handbook along with our own personal experiences. Many people have shaped how we go about doing ecology, and we have borrowed heavily from what we have been taught formally and informally. We thank Anurag Agrawal, Winnie Anderson, Jim Archie, Shelley Berc, Leon Blaustein, Gideon Bradburd, Liz Constable, the CPSI group at Buffalo State University, the d.school group at Stanford University, Will Davis, Teresa Dillinger, Hugh Dingle, Alejandro Fogel, Jeff Granett, Patrick Grof-Tisza, Jessica Gurevitch, Marcel Holyoak, Henry Horn, David Hougen-Eitzman, Apryl Huntzinger, Erika Iyengar, Dan Janzen, Claire Karban, Sharon Lawler, Rich Levine, Monte Lloyd, Greg Loeb, Eric LoPresti, John Maron, Toshiko Murata, Naomi Murray, Rob Page, Adam Pepi, Sanjay Pyare, Jim Quinn, Dave Reznick, Kevin Rice, Bob Ricklefs, Danielle Rutkowski, Tom Scott, Tina Seelig, Kaori Shiojiri, Jonathan Shurin, Andy Sih, Chris Simon, Dean Keith Simonton, Sharon Strauss, Don Strong, Jennifer Thaler, Will Wetzel, Neil Willets, Louie Yang, and Truman Young, all of whom made valuable contributions to this book. Many people generously provided feedback on the first and second editions that has improved this third one. We thank Christer Björkman, Erika Iyengar, Sarah Kimball, Don Miles, Christopher Norment, Jennifer Peterson, Heather Throop, Neal

Williams, and especially, Truman Young, for helpful critique. We thank Dehua Wang for translating the first two editions into Mandarin. We are fairly certain that we have neglected to mention many others, and we apologize for these unintentional omissions. We are also grateful to Alison Kalett for really getting our vision for this book and supporting us through the first two editions, to Jodi Beder for unusually insightful editing and improving all three editions, and to Sophia Zengierski and Sydney Carroll for tirelessly showering us with help as we worked on the third edition—thank you all for making the writing of this book so much fun.

References

Acevedo, F. E., L. J. Rivera-Vega, S. H. Chung, S. Ray, and G. W. Felton. 2015. Cues from chewing insects: The intersection of DAMPs, HAMPs, MAMPs and effectors. Current Opinion in Plant Biology 26:80–86.

Agrawal, A. A., and P. M. Kotanen. 2003. Herbivores and the success of exotic plants: A phylogenetically controlled experiment. Ecology Letters 6:712–15.

Ajani, K., E. Lee, C. Xiong, C. N. Knaflic, W. Kemper, and S. Franconeri. 2021. Declutter and focus: Empirically evaluating design guidelines for effective data communication. IEEE Transactions on Visualization and Computer Graphics, doi: 10.1109 /TVCG.2021.3068337.

Anderson, J. G. T. 2017. Why ecology needs natural history. American Scientist 205:290–98.

Arnqvist, G. 2020. Mixed models offer no freedom from degrees of freedom. Trends in Ecology and Evolution 35:329–35.

Aschwanden, C. 2015. Science isn't broken. https://fivethirtyeight .com/features/science-isnt-broken/.

Austin, A. E., and E. R. Miller. 2020. Strengths, challenges and opportunities for US doctoral education. In M. Yudkevich, P. G. Altbach, and H. de Wit, eds., *Trends and Issues in Doctoral Education: A Global Perspective.* Thousand Oaks, CA: SAGE.

Baas, M., C. K. W. De Dreu, and B. A. Nijstad. 2008. A meta-analysis of 25 years of mood-creativity research: Hedonic tone, activation, or regulatory focus? Psychological Bulletin, 134:779–806.

Baldwin, I. T. 1988. The alkaloidal responses of wild tobacco to real and simulated herbivory. Oecologia 77:378–81.

Beck, M. 2001. Finding Your Own North Star. New York: Three Rivers Press.

Bergerud, A. T., and W. E. Mercer. 1989. Caribou introductions in eastern North America. Wildlife Society Bulletin 17:111–20.

Berner, D., and V. Amrhein. 2022. Why and how we should join the shift from significance testing to estimation. Journal of Evolutionary Biology 35:777–87.

Bertness, M. D., and R. M. Callaway. 1994. Positive interactions in communities. Trends in Ecology and Evolution 9:191–93.

Blickley, J. L., K. Deiner, K. Garbach, I. Lacher, M. H. Meek, L. M. Porensky, M. L. Wilkerson, E. M. Winford, and M. W. Schwartz. 2013. Graduate student's guide to necessary skills for nonacademic conservation careers. Conservation Biology 27:24–34.

Bolles, R. N., and K. Brooks. 2020. What Color Is Your Parachute? New York: Ten Speed Press.

Boot N., M. Baas, E. Mühlfeld, C. K. de Dreu, and S. van Gaal. 2017. Widespread neural oscillations in the delta band dissociate rule convergence from rule divergence during creative idea generation. Neuropsychologia 104:8–17.

Boudreau, K. J., E. C. Guinan, K. R. Lakhani, and C. Riedl. 2016. Looking across and looking beyond the knowledge frontier: Intellectual distance, novelty, and resource allocation in science. Management Science 62:2765–83.

Bowman, J., J. C. Ray, A. J. Magoun, D. S. Johnson, and F. N. Dawson. 2010. Roads, logging, and the large-mammal community of an eastern Canadian boreal forest. Canadian Journal of Zoology 88:454–67.

Bricchi, I., M. Leitner, M. Foti, A. Mithofer, W. Boland, and M. E. Maffei. 2010. Robotic mechanical wounding (MecWorm) versus herbivore-induced responses: early signaling and volatile emission in Lima bean (*Phaseolus lunatus* L.). Planta 232: 719–29.

Brice, E. M., E. J. Larsen, and D. R. MacNulty. 2022. Sampling bias exaggerates a textbook example of a trophic cascade. Ecology Letters 25:177–88.

Brown, J. H., and M. V. Lomolino. 1989. Independent discovery of the equilibrium theory of island biogeography. Ecology 70:1955–57.

Burnham, K. P., and D. R. Anderson. 2002. Model selection and multimodel inference: a practical information-theoretic approach. 2nd ed. New York: Springer Verlag.

Calcagno, V., E. Demoinet, K. Gollner, L. Guidi, D. Ruths, and C. de Mazancourt. 2012. Flows of research manuscripts among scientific journals reveal hidden submission patterns. Science 338: 1065–69.

Cassey, P., and T. M. Blackburn. 2004. Publication and rejection among successful ecologists. BioScience 54:234–39.

Chandler, C. R., L. M. Wolfe, and D. E. L. Promislow. 2007. The Chicago Guide to Landing a Job in Academic Biology. Chicago: University of Chicago Press.

Coelho, D. A., and F. L. Vieira. 2018. The effect of previous group interaction on individual ideation novelty and variety. International Journal of Design Creativity and Innovation 6:80–92.

Cohen, J. 1988. Statistical Power Analysis for the Behavioral Sciences. 2nd ed. Hillsdale, NJ: Lawrence Erlbaum.

Cottingham, K. L., J. T. Lennon, and B. L. Brown. 2005. Knowing when to draw the line: designing more informative ecological experiments. Frontiers in Ecology and the Environment 3:145–52.

Crilly, N. 2015. Fixation and creativity in concept development: The attitudes and practices of expert designers. Design Studies 38:54–91.

Cropley, A. 2006. In praise of convergent thinking. Creativity Research Journal 18:391–404.

Crouse, D. T., L. B. Crowder, and H. Caswell. 1987. A stage-based population model for loggerhead sea turtles and implications for conservation. Ecology 68:1412–23.

Damgaard, C. 2019. A critique of the space-for-time substitution practice in community ecology. Trends in Ecology and Evolution 34:416–21.

Damrosch, D. 1995. We Scholars: Changing the Culture of the University. Cambridge, MA: Harvard University Press.

Darwin, C. 1889. The Origin of Species. 6th ed. New York: D. Appleton.

Deegan, D. H. 1995. Exploring individual differences among novices reading in a specific domain: the case of law. Reading Research Quarterly 30:154–57.

Diamond, J. 1986. Overview: Laboratory experiments, field experiments, and natural experiments. Pages 3–22 in J. Diamond and T. J. Case, eds., Community Ecology. New York: Harper and Row.

Dray, S., R. Pelissier, P. Couteron, M. J. Fortin, P. Legendre, P. R. Peres-Neto, E. Bellier, R. Bivand, F. G. Blanchet, M. De Caceres, A. G. Dufour, E. Heegaard, T. Jombart, F. Munoz, J. Oksanen, J. Thioulouse, and H. H. Wagner. 2012. Community ecology in the age of multivariate multiscale spatial analysis. Ecological Monographs 82:257–75.

Duchardt, C. J., L. M. Porensky, and I. S. Pearse. 2021. Direct and indirect effects of a keystone engineer on a shrubland-prairie food web. Ecology 102:e03195.

Dugosh, K. L., P. B. Paulos, E. J. Roland, and H. C. Yang. 2000. Cognitive stimulation in brainstorming. Journal of Personality and Social Psychology 79:722–35.

Edmondson, A. C., and Z. Lei. 2014. Psychological safety: the history, renaissance, and future of an interpersonal construct. Annual Review of Organizational Psychology and Organizational Behavior 1:23–43.

Eidelman, S., C. S. Crandall, and J. Pattershall. 2009. The existence bias. Journal of Personality and Social Psychology 97:765–75.

Ellamil, M., C. Dobson, M. Beeman, and K. Christoff. 2012. Evaluative and generative modes of thought during the creative process. Neuroimage 59:1783–94.

Feist, G. J. 1998. A meta-analysis of personality in scientific and artistic creativity. Personality and Social Psychology Review 2:290–309.

Fick, S. E., T. W. Nauman, C. C. Brungard, and M. C. Duniway. 2021. Evaluating natural experiments in ecology: Using synthetic controls in assessments of remotely sensed land treatments. Ecological Applications 31:e02264.

Finkbeiner, E. M., B. P. Wallace, J. E. Moore, R. L. Lewison, L. B. Crowder, and A. J. Read. 2011. Cumulative estimates of sea turtle bycatch and mortality in USA fisheries between 1990 and 2007. Biological Conservation 144:2719–27.

Finke, R., T. B. Ward, and S. M. Smith. 1992. Creative Cognition: Theory, Research, and Applications. Cambridge, MA: MIT Press.

Foucault, M. 1977. Discipline and Punish: The Birth of the Prison. New York: Pantheon.

Fox, J. 2020. A data-based guide to the North American ecology faculty job market. Bulletin of the Ecological Society of America 101:e01624.

Friedland, A. J., C. L. Folt, and J. L. Mercer. 2018. Writing successful science proposals. 3rd ed. New Haven: Yale University Press.

Gana, K., and G. Broc. 2019. Structural equation modeling with *lavaan*. Hoboken, NJ: Wiley.

Garland, T., A. F. Bennett, and E. L. Rezende. 2005. Phylogenetic approaches in comparative physiology. Journal of Experimental Biology 208:3015–35.

Girotra, K., C. Terwiesch, and K. T. Ulrich. 2010. Idea generation and the quality of the best idea. Management Science 56:591–605.

Golde, C. M., and T. M. Dore. 2001. At cross purposes: What the experiences of doctoral students reveal about doctoral education. A report prepared for the Pew Charitable Trusts, Philadelphia. www.phd-survey.org.

Gotelli, N. J., and A. M. Ellison. 2004. A Primer of Ecological Statistics. Sunderland, MA: Sinauer.

Grace, J. B. 2006. Structural Equation Modeling and Natural Systems. Cambridge: Cambridge University Press.

Grof-Tisza, P., M. Holyoak, E. Antell, and R. Karban. 2015. Predation and associational refuge drive ontogenetic niche shifts in an arctiid caterpillar. Ecology 96:80–89.

Hale, A. N., S. J. Tonsor, and S. Kalisz. 2011. Testing the mutualism disruption hypothesis: physiological mechanisms for invasion of intact perennial plant communities. Ecosphere 2:110.

Hampton, S. E., and S. G. Labou. 2017. Careers in ecology: a fine-scale investigation of national data from the U.S. Survey of Doctorate Recipients. Ecosphere 8:e02031.

Hilborn, R., and M. Mangel. 1997. The Ecological Detective. Confronting Models with Data. Princeton, NJ: Princeton University Press.

Hofer, T., H. Przyrembel, and S. Verleger. 2004. New evidence for the theory of the stork. Paediatric and Perinatal Epidemiology 18:88–92.

Holt, R. D. 1977. Predation, apparent competition and the structure of prey communities. Theoretical Population Biology 12:197–229.

Holt, R. D., and M. B. Bonsall. 2017. Apparent competition. Annual Review of Ecology, Evolution, and Systematics 48:447–71.

Huberty, A. F., and R. F. Denno. 2004. Plant water stress and its consequences for herbivorous insects: A new synthesis. Ecology 85:1383–98.

Huntzinger, M. 2003. Effects of fire management practices on butterfly diversity in the forested western United States. Biological Conservation 113:1–12.

Huntzinger, M., R. Karban, T. P. Young, and T. M. Palmer. 2004. Relaxation of induced indirect defenses of acacias following exclusion of mammalian herbivores. Ecology 85:609–14.

Hurlbert, S. H. 1984. Pseudoreplication and the design of ecological field experiments. Ecological Monographs 54:187–211.

Janes, L. M., and J. M. Olson. 2010. Is it you or is it me? Contrasting effects of ridicule targeting other people versus the self. Europe's Journal of Psychology 6:46–70.

Janes, L. M., and J. M. Olson. 2015. Humor as an abrasive or a lubricant in social situations: Martineau revisited. Humor 28:271–88.

Jauk, E., M. Benedek, and A. C. Neubauer. 2012. Tackling creativity at its roots: Evidence for different patterns of EEG alpha activity related to convergent and divergent modes of task processing. International Journal of Psychophysiology 84:219–25.

Johnson, B. R., and C. J. D'Lauro. 2018. After brainstorming, groups select an early generated idea as their best idea. Small Group Research 49:177–94.

Jung, R. E., C. J. Wertz, C. A. Meadows, S. G. Ryman, A. A. Vakhtin, and R. A. Flores. 2015. Quantity yields quality when it comes to creativity: A brain and behavioral test of the equal-odds rule. Frontiers in Psychology 25:864.

Karban, R. 1983. Induced responses of cherry trees to periodical cicada oviposition. Oecologia 59:226–31.

Karban, R. 1987. Environmental conditions affecting the strength of induced resistance against mites in cotton. Oecologia 73:414–19.

Karban, R. 1989. Community organization of *Erigeron glaucus* folivores: effects of competition, predation, and host plant. Ecology 70:1028–39.

Karban, R. 1993. Costs and benefits of induced resistance and plant density for a native shrub, *Gossypium thurberi*. Ecology 74:9–19.

Karban, R., and I. T. Baldwin. 1997. Induced Responses to Herbivory. Chicago: University of Chicago Press.

Karban, R., and P. de Valpine. 2010. Population dynamics of an arctiid caterpillar-tachinid parasitoid system using state-space models. Journal of Animal Ecology 79:650–61.

Karban, R., and J. Maron. 2001. The fitness consequences of interspecific eavesdropping between plants. Ecology 83:1209–13.

Karban, R., T. Mata, P. Grof-Tisza, G. Crutsinger, and M. Holyoak. 2013. Non-trophic effects of litter reduce ant predation and determine caterpillar survival and distribution. Oikos 122:1362–70.

Kassim, A. A., R. Rehman, and J. M. Price. 2018. Effects of modality and repetition in a continuous recognition memory task: Repetition has no effect on auditory recognition memory. Acta Psychologica 185:72–80.

Koenig, W. D., and J. M. H. Knops. 2013. Large-scale spatial synchrony and cross-synchrony in acorn production by two California oaks. Ecology 94:83–93.

Koenig, W. D., J. M. H. Knops, and W. J. Carmen. 2010. Testing the environmental prediction hypothesis for mast-seeding in California oaks. Canadian Journal of Forest Research 40:2115–22.

Koenig, W. D., J. M. H. Knops, W. J. Carmen, M. T. Stanback, and R. L. Mumme. 1996. Acorn production by oaks in central coastal California: Influence of weather at three levels. Canadian Journal of Forest Research–Revue Canadienne de Recherche Forestiere 26:1677–83.

Kohn, N. W., P. B. Paulus, and Y. Choi. 2011. Building on the ideas of others: An examination of the idea combination process. Journal of Experimental Social Psychology 47:554–61.

Konno, K. 2016. A general parameterized mathematical food web model that predicts a stable green world in the terrestrial ecosystem. Ecological Monographs 86:190–214.

Koricheva, J., J. Gurevitch, and K. Mengersen. 2013. Handbook of Meta-analysis in Ecology and Evolution. Princeton, NJ: Princeton University Press.

Legendre, P., M. R. T. Dale, M.-J. Fortin, F. Gurevitch, M. Hohn, and D. Myers. 2002. The consequences of spatial structure for the design and analysis of ecological field surveys. Ecography 25:601–15.

Lertzman, K. 1995. Notes on writing papers and theses. Bulletin of the Ecological Society of America. June1995:86–90.

MacArthur, R. H., and E. O. Wilson. 1963. An equilibrium theory of insular zoogeography. Evolution 17:373–87.

MacArthur, R. H., and E. O. Wilson. 1967. The Theory of Island Biogeography. Princeton, NJ: Princeton University Press.

Madjar, N., E. Greenberg, and Z. Chen. 2011. Factors for radical creativity, incremental creativity, and routine, noncreative performance. Journal of Applied Psychology 96:730–43.

Marguc, J., G. A. Van Kleef, and J. Förster. 2015. Welcome interferences: Dealing with obstacles promotes creative thought in goal pursuit. Creativity and Innovation Management 24:207–16.

Maron, J. L., and S. Harrison. 1997. Spatial pattern formation in an insect host-parasitoid system. Science 278:1619–21.

Marquis, R. J., and C. J. Whelan. 1995. Insectivorous birds increase growth of white oak through consumption of leaf-chewing insects. Ecology 75:2007–14.

Martin, T. E. 1993. Nest predation and nest sites: New perspectives on old patterns. BioScience 43:523–32.

Matthews, R. 2000. Storks deliver babies ($p=0.008$). Teaching Statistics 22:36–38.

Mayseless, N., J. Aharon-Peretz, and S. Shamay-Tsoory. 2014. Unleashing creativity: The role of left temporoparietal regions in evaluating and inhibiting the generation of creative ideas. Neuropsychologia 64:157–68.

McElreath, R. 2019. Statistical Rethinking: A Bayesian course with examples in R and Stan. 2nd ed. Boca Raton, FL: CRC Press.

Mooney, K. A., D. S. Gruner, N. A. Barber, S. A. Van Bael, S. M. Philpott, and R. Greenberg. 2010. Interactions among predators and the cascading effects of vertebrate insectivores on arthropod communities and plants. Proceedings of the National Academy of Sciences 107:7335–40.

Mueller, J. S., S. Melwani, and J. A. Goncalo. 2012. The bias against creativity: Why people desire but reject creative ideas. Psychological Science 23:13–17.

Mueller J.S., C. J. Wakslak, and V. Krishnan. 2014. Construing creativity: The how and why of recognizing creative ideas. Journal of Experimental Social Psychology 51:81–87.

Muñoz Adánez, A. 2005. Does quantity generate quality? Testing the fundamental principle of brainstorming. Spanish Journal of Psychology 8:215–20.

Munroe, E. G. 1948. The geographical distribution of butterflies in the West Indies. Dissertation, Cornell University.

Munroe, E. G. 1953. The size of island faunas. Pages 52–53 in Proceedings of the Seventh Pacific Science Congress of the Pacific Science Association, vol. 4: Zoology. Auckland: Whitcome and Tombs.

Newmark, W. D. 1995. Extinction of mammal populations in western North American national parks. Conservation Biology 9:512–26.

Newmark, W. D. 1996. Insularization of Tanzanian parks and the local extinction of large mammals. Conservation Biology 10:1549–56.

Nijstad, B. A., and W. Stroebe. 2006. How the group affects the mind: A cognitive model of idea generation in groups. Personality and Social Psychology Review 10:186–213.

Oksanen, L. 2001. Logic of experiments in ecology: Is pseudoreplication a pseudoissue? Oikos 94:27–38.

Oppezzo, M., and D. L. Schwartz. 2014. Give your ideas some legs: The positive effect of walking on creative thinking. Journal of

Experimental Psychology: Learning, Memory, and Cognition 40:1142.

Osborn, A. 1953. Applied Imagination: Principles and Procedures of Creative Problem Solving. New York: Scribner.

Paulus, P. B., N. W. Kohn, and L. E. Arditti. 2011. Effects of quantity and quality instructions on brainstorming. Journal of Creative Behavior 45:38–46.

Pearse, I. S. 2011. The role of leaf defensive traits in oaks on the preference and performance of a polyphagous herbivore, *Orgyia vetusta*. Ecological Entomology 36:635–42.

Pearse, I. S., and A. L. Hipp. 2009. Phylogenetic and trait similarity to a native species predict herbivory on non-native oaks. Proceedings National Academy of Sciences 106:18097–102.

Pearse, I. S., R. Paul, and P. J. Ode. 2018. Variation in plant defense suppresses herbivore performance. Current Biology 28:1981–86.

Peay, K. G., M. Belisle, and T. Fukami. 2012. Phylogenetic relatedness predicts priority effects in nectar yeast communities. Proceedings of the Royal Society B 279:749–58.

Pechmann, J. H. K., D. E. Scott, R. D. Semlitsch, J. P. Caldwell, L. J. Vitt, and J. W. Gibbons. 1991. Declining amphibian populations: The problem of separating human impacts from natural fluctuations. Science 253:892–95.

Pepi, A., V. Pan, D. Rutkowski, V. Mase, and R. Karban. 2022. Influence of delayed density and ultraviolet radiation on caterpillar baculovirus infection and mortality. Journal of Animal Ecology 91:2192–2202.

Platt, J. R. 1964. Strong inference. Science 146:347–53.

Popper, K. R. 1959. The Logic of Scientific Discovery. New York: Basic Books.

Potvin, C. 1993. ANOVA: Experiments in controlled environments. Pages 46–68 in S. M. Scheiner and J. Gurevitch, eds., Design and Analysis of Ecological Experiments. New York: Chapman and Hall.

Quinn, J. F., and A. E. Dunham. 1983. On hypothesis testing in ecology and evolution. American Naturalist 122:602–17.

Reznick, D. N., and J. A. Endler. 1982. The impact of predation on life history evolution in Trinidadian guppies (*Poecilia reticulata*). Evolution 36:160–77.

Reznick, D. N., H. Bryga, and J. A. Endler. 1990. Experimentally induced life-history evolution in a natural population. Nature 346:357–59.

Ricklefs, R. E. 2012. Naturalists, natural history, and the nature of biological diversity. American Naturalist 179:423–35.

Ricklefs, R. E., and D. Schluter. 1993. Species diversity: Regional and historical influences. Pages 350–63 in R. E. Ricklefs and D. Schluter, eds., Species Diversity in Ecological Communities. Chicago: University of Chicago Press.

Rietzschel, E. F., B. A. Nijstad, and W. Stroebe. 2006. Productivity is not enough: A comparison of interactive and nominal brainstorming groups on idea generation and selection. Journal of Experimental Social Psychology 42:244–51.

Rietzschel, E. F., B. A. Nijstad, and W. Stroebe. 2010. The selection of creative ideas after individual idea generation: Choosing between creativity and impact. British Journal of Psychology 101:47–68.

Roche, M. D., I. S. Pearse, L. Bialic-Murphy, S. N. Kivlin, H. R. Sofaer, and S. Kalisz. 2021. Negative effects of an allelopathic invader on AM fungal plant species drive community-level responses. Ecology 102:e03201.

Roche, M. D., I. S. Pearse, H. R. Sofaer, S. N. Kivlin, G. Spyreas, D. N. Zaya, and S. Kalisz. Forthcoming. Invasion-mediated mutualism disruption is evident across heterogeneous environmental conditions and varying invasion intensities. Ecography.

Rosso, B. D. 2014. Creativity and constraints: Exploring the role of constraints in the creative processes of research and development teams. Organization Studies 35:551–85.

Runco, M. A. 2020. Tactics and strategies. Pages 529–32 in M. A. Runco and S. Pritzker, eds., Encyclopedia of Creativity, 3rd ed. N.p.: Academic Press.

Schimel, J. 2012. Writing science: How to write papers that get cited and proposals that get funded. New York: Oxford University Press.

Schneider, D. C., R. Walters, S. Thrush, and P. Dayton. 1997. Scale-up of ecological experiments: Density variation in the mobile bivalve Macomona liliana. Journal of Experimental Marine Biology and Ecology 216:129–52.

Schwartz, M. W., J. K. Hiers, F. W. Davis, G. M. Garfin, S. T. Jackson, A. J. Terando, C. A. Woodhouse, T. L. Morelli, M. A. Williamson, and M. W. Brunson. 2017. Developing a translational ecology workforce. Frontiers in Ecology and the Environment 15:587–96.

Smith, L. L., A. L. Subalusky, C. L. Atkinson, J. E., Earl, D. M. Mushet, D. E. Scott, S. L. Vance, and S. A. Johnson. 2019. Biological connectivity of seasonally ponded wetlands across spatial and temporal scales. Journal of the American Water Resources Association 55:334–53.

Shipley, B. 2000. Cause and Correlation in Biology: A User's Guide to Path Analysis, Structural Equations and Causal Inference. Cambridge: Cambridge University Press.

Shurin, J. B., E. T. Borer, E. W. Seabloom, K. Anderson, C. A. Blanchette, B. Broitman, S. D. Cooper, and B. S. Halpern. 2002. A cross-ecosystem comparison of the strength of trophic cascades. Ecology Letters 5:785–91.

Sih, A., J. Cote, M. Evans, S. Fogarty, and J. Pruitt. 2012. Ecological implications of behavioral syndromes. Ecology Letters 15:278–89.

Silvia, P. J., E. C. Nusbaum, and R. E. Beaty. 2017. Old or new? Evaluating the old/new scoring method for divergent thinking tasks. Journal of Creative Behavior 51:216–24.

Simonton, D. K. 2003. Scientific creativity as constrained stochastic behavior: The integration of product, person, and process perspectives. Psychological Bulletin 129:475.

Sokal, R. R., and F. J. Rohlf. 2012. Biometry. 4th ed. New York: Freeman.

Thomas, D. C., and D. R. Gray. 2002. Update COSEWIC status report on the woodland caribou *Rangifer tarandus caribou* in Canada. *In COSEWIC assessment and update status report on the woodland caribou* Rangifer tarandus caribou *in Canada*. Ottawa: Committee on the Status of Endangered Wildlife in Canada.

Todd, B. T., D. E. Scott, J. H. K. Pechmann, and J. W. Gibbons. 2011. Climate change correlates with rapid delays and advancements in reproductive timing in an amphibian community. Proceedings of the Royal Society B 278:2191–97.

Torrance, E. P., and J. P. Torrance. 1978. Developing creativity instructional materials according to the Osborn-Parnes creative problem solving model. Creative Child and Adult Quarterly 3:80–90.

Valdiva, A., S. Wolf, and K. Suckling. 2019. Marine mammals and sea turtles listed under the U.S. Endangered Species Act are recovering. PLoS One 14:e0210164

van Kleunen, M., et al. 2015. Global exchange and accumulation of non-native plants. Nature 525:100–103.

Vannette, R. L., M. P. L. Gauthier, and T. Fukami. 2013. Nectar bacteria, but not yeast, weaken a plant-pollinator mutualism. Proceeding of the Royal Society B 280:1752.

Vaughn, K. J., and T. P. Young. 2010. Contingent conclusions: Year of initiation influences ecological field experiments, but temporal replication is rare. Restoration Ecology 18:59–64.

Vitt, L. J., and E. R. Pianka. 2005. Deep history impacts present-day ecology and biodiversity. Proceedings of the National Academy of Sciences 102:7877–81.

Vors, L. S., and M. S. Boyce. 2009. Global declines of caribou and reindeer. Global Change Biology 15:2626–33.

Weber, M. G., and A. A. Agrawal. 2012. Phylogeny, ecology, and the coupling of comparative and experimental approaches. Trends in Ecology and Evolution 27:394–403.

Wetzel, W. C., H. M. Kharouba, M. Robinson, M. Holyoak, and R. Karban. 2016. Variability in plant nutrients reduces insect herbivore performance. Nature 539:425–27.

White, T. C. R. 1969. An index to measure weather-induced stress of trees associated with outbreaks of psyllids in Australia. Ecology 50:905–9.

White, T. C. R. 1984. The abundance of invertebrate herbivores in relation to the availability of nitrogen in stressed food plants. Oecologia 63:90–105.

White, T. C. R. 2008. The role of food, weather and climate in limiting the abundance of animals. Biological Reviews 83:227–48.

Yang, L. H. 2004. Periodical cicadas as resource pulses in North American forests. Science 306:1565–67.

Yoccuz, N. G. 1991. Use, overuse, and misuse of significance tests in evolutionary biology and ecology. Bulletin of the Ecological Society of America 72:106–11.

Young, T. P., B. Okello, D. Kinyua, and T. M. Palmer. 1998. KLEE: A long-term, large-scale herbivore exclusion experiment in Laikipia, Kenya. African Journal of Range and Forage Science 14:94–102.

Zhu Y., S. M. Ritter, and A. Dijksterhuis. 2020. Creativity: Intrapersonal and interpersonal selection of creative ideas. Journal of Creative Behavior 54:626–35.

Zschokke, S., and E. Ludin. 2001. Measurement accuracy: How much is necessary? Bulletin of the Ecological Society of America 82:237–43.

Index

abstract: journal, 142–143; posters, 175; proposal, 182–183; writing process, 138

accuracy, 43

acorns, 81

advisor. *See* professor, major

Akaike Information Criteria (AIC), 91

analysis: least squares, 54, 57, 90; phylogenetically explicit, 57; spatially explicit, 54

analysis of variance (ANOVA), 70–71

antler shedding, 49

appendices, 136

application: grad school, 1; job, 117, 126–131

aquariums, 58–59

article, journal, 142–154; publishing process, 154, 157–158

artifacts. *See* experiments: artifacts

assistants, being one, 117; hiring one, 123–124; teaching, 127, 130

assumptions, 8, 24, 29

attention, at seminars, 108, 161, 162

audience, of oral presentations, 141, 144, 158–174; of journal articles, 141, 144; of posters, 172, 175

authorship, 118–119

autocorrelation. *See* replication: non-independence

axes, figure, 150; in oral presentations, 150, 165

babies: where they come from, 65–66

Bayesian Information Criteria (BIC), 91

Bayesian statistics. *See* statistics: Bayesian

Beck, Martha, 125–126

bias, 43–44, 64; in idea generation, 100

biology-speak, 150

birth rate, human, 65–66

blocking, 53, 110

brainstorming, 96–98, 101–105

budget, 181, 182

cage effects, 38–39

car color, 31

career, 15–16, 83, 120; non-academic, 132; teaching, 130

caribou, 26–27, 77–80

causation, 31–33, 38, 58–59; manipulative experiments, 21, 23, 29–30, 33, 38, 41, 76, 190; path analysis, 77–80, 91; quantitative observations, 22–23, 77–80, 87–88, 89–90; replication, 41–42; structural equation modeling, 33, 91–92; vs correlation, 29, 30, 33, 89–90

checklist: journal article, 155–157; oral presentation, 173–175; poster, 178–179; proposals, 186–187

cicadas, 12, 36

citations, 154

coauthors, 118–119, 121, 175

coevolution, 8
collaborators, 29, 119, 121, 122, 134
collinearity, 89–90
colorblindness, 162, 175
committees, 119–120, 134, 172
communication, 117, 123, 134, 155, 173, 178, 186; value of, 15, 135–136, 137
community colleges, 126, 130
competition, 22; apparent, 25–27
conclusion: in papers, 153–154, 157; on posters, 176, 178; in talks, 173
conferences, 121–122; posters at, 172, 175–179
confounding factors. *See* factors, confounding
consequences, 11–12
conservation, 25, 44
constraints, 17, 100, 104, 111
controlled environment, 58–60
control, social, 98
controls, 38–39, 41, 49, 51, 191; quantitative observations, 21; manipulative experiments, 23
co-occurrence, 22
correlation, 29–34, 89–90
cost, of rules, 190–192
counterexample, 67
covariation, 86, 90
creativity: research on, 96–100; workshop, 101–105
critique, 102, 107–108
currency, xiii-xiv, 192
curriculum vitae (CV), 129

Darwin, Charles, 28, 113, 135
data: analysis, 71–72, 112; big, 91, 95; communicating your, 149, 151, 152, 160, 165; extrapolating, 81; flawed, 94; long-term, 22, 83–84;

observational, 23, 33, 77; preliminary, 10, 36, 85–87, 182; sheets, 110; whaling, 35
deer, 26–27, 77–80
discussion: in papers, 152–153, 157; in posters, 178; in proposals, 182; in talks, 173
diversity (DEI), 131–132
draft: perfect, 138, 140; rough, 112, 140, 141

effect: cause and, 30–31, 38–41, 49, 50, 65; collinear, 89, 91; covariate, 86; fixed, 54–55; indirect, 64, 79–80; linear, 70–71, 90; non-linear, 70–71, 90; random, 54–55; shared ancestry, 57; size, 44, 61–62, 67, 73–74, 78–80, 149–150, 156, 190
English as a non-native language, 142, 157, 163, 168–170
equations, and oral presentations, 164
equity (DEI), 131–132
error: bars, 151; standard, 62, 151; statistical, 65; type I, 65; type II, 65
evaluating ideas, 96–100; criticism, 102
exam, qualifying, 119; defensiveness, 172
experiments: artifacts, 38–39, 60; manipulative, 17, 21, 23–24, 30, 33, 34, 35–60; natural, 22, 60, 76; pilot, 3; scale, 21–22, 34, 44–45, 46–48, 60
extinction, 48
extrapolation, 21, 47–48, 74, 81, 153

face time, 118
factors: 18–19, 29, 58, 69, 70, 89; causal, 24, 29, 32, 89, 91; confounding, 38, 53, 57, 83; figures,

149; multiple, 73, 77, 87–89, 91; number of, 18–19, 41–45, 93; which to include, 18–19, 24, 77, 84–87
failure, 4, 110
feasibility, 4, 109, 181, 182, 187
feedback, 109, 110, 169; posters, 177
fellowships, 116
field stations, 8
fieldwork, 110, 128; for grants, 182
figures: axes, 150, 165; captions, 150; color-blind friendly, 162; in journal articles, 149–151, 156; in oral presentations, 162; on posters, 176–177; to represent relationships, 20, 62, 92, 146, 149. *See also* graphs
fire, as a treatment, 40
fitness, 8
format: journal, 142; oral presentation, 159–160; poster, 175–177; proposal, 182
funding: and authorship, 119; constraints, 2, 3, 45; and grad school, 116; grants, 179–182, 187–188

generality, 2, 3, 86; big data, 94–95; in discussion, 152–153, 176; in intro, 143–144, 146, 167, 184; meta-analyses, 74; modeling, 24
generating ideas. *See* idea generation
goals, 3, 14–16, 105, 116, 192; job, 14, 126; writing process, 139
gradient, environmental, 53
graduate school, xiii, 9, 115–119; applying, 116–119; getting in, 1; messing with your head, 125; reading, 106–108; transferrable skills, 127–128
grants. *See* proposals

graphs: bar, 151; directed, 32–33, 92; scatterplots, 151, 156. *See also* figures
greenhouse, 50–51, 54–55
growth chambers, 58
guppies, 42

harbor, 46–47
herbivore exclusion, 38–39, 44
herbivory, 12, 56
heterogeneity, environmental, 53
history, evolutionary, 8, 55–57
house of cards, 68
hypotheses: alternative, 67–69, 70, 85, 100, 101; generating, 17, 19, 25, 77, 190; null, 63–67, 71–72; in papers, 146, 148; in proposals, 184; SEM, 91–93; testing, 15, 21, 22, 24, 28, 35–60, 61; yes/no, 70

Ian, world of, 57, 59, 88–89, 132, 133, 189
idea generation, 96–100, 113; workshop for, 101–105
images, for communicating, 122, 161, 177
impact factors: of journals, 136
imprecision, 43
inclusion (DEI), 131–132
independence, 41–42, 44, 50–52, 84, 87, 92, 110; phylogenetic, 55–57; spatial, 53–55
induced resistance, 59–60
information criteria, 91
information management, 20
insect outbreaks, 32
internship, 134
interspersion, 22, 50–53, 93, 191
interview, 117–119, 129

introduction: journal, 143–147; proposal, 180, 182–183
intuition, 19, 34
island biogeography, theory of, 136

Jesse, world of, 67–68
jobs, xiii, 5, 14–16, 115, 118, 120, 124–134; creating your own, 129; listings, 128, 134
journal article. *See* article, journal
justification, in proposals, 180, 183, 184, 186

lab: artifacts, 58–59; experimental design, 50–55, 58–60; mates, 19, 109, 120, 169; meetings, 98, 120
laser pointer, 170
leprechaun trap, 67
letter: cover, 117, 127, 130, 158; from editor, 158
levels. *See* treatment, levels
lights, during oral presentations, 170
linear effects. See effects, linear
literature, 106–108, 112, 145–146, 160; review, 112
local determinism, 48

MacArthur, Robert, 23, 136
major professor. *See* professor, major
Mao, the game, 190
means, 43, 62–63, 73–74, 111, 149, 151; controlled conditions, 59; vs individual variation, 12
mechanisms, 11–12, 28, 58, 68, 77; alternative, 28, 68; model, 28–29
meetings. *See* conferences *and* lab, meetings
mentoring, 117–120, 124, 125
message, take home. *See* take-home message

meta-analysis, 72–75
methods, 111, 191; in articles, 138, 147–148, 155; on posters, 176; in proposals, 182; in talks, 159, 161, 164
microcosm, 58
microenvironment, 38–39
Mikaela, world of, 19, 44, 100, 125, 139, 165, 166, 169
models: building, 24–29, 33; mathematical, 24, 146, 163; mixed, 54–55; testing, 28

natural experiments. *See* experiments, natural
natural history, 17–18, 191; in articles, 146–147
negative results, 63, 72–74
nervousness, and oral presentations, 166, 171
noise, 41, 44, 45; blocking, 53; controlled environments, 58–59; error bars, 151
non-linear effects. *See* effects, non-linear
non-profit government organization (NGO), 120, 127, 133, 134
notebook, 20, 111
notes, in oral presentations, 166
novelty, 1–2, 104, 154, 180, 181

oaks, 57, 81, 88
objectives, of proposals, 182–186
objectivity, 115
observations, 17–23, 30, 34, 35, 42, 54, 76–95, 109, 111, 182, 185, 191; analysis, 77, 84, 87–93; better than manipulations, 21; causality, 22, 30; study requisites, 37
opportunity, 28, 83, 117, 122, 127, 130, 191

oral presentations: checklist, 173–175; giving, 141, 158–172, question-and-answer (Q&A), 167–168, 169, 171–172

organization: 138–140; articles, 142–143, 148; posters, 176–177, 179; proposals, 182–185; talks, 158–161, 163–165

outline, 138–140; in oral presentations, 159; slide, 164–165

p-value, 61–66

panopticon, 98–99

parents, 16, 98

party trick, 43

path, 110, 160; analysis, 77–80, 91–92; diagram, 33, 77–80, 91, 146

pattern: alternative hypotheses, 67, 69; mechanisms and consequences, 11–12; modeling, 24, 28, 34; observing and quantifying, 10–11, 17, 20–21, 23, 58, 61, 76–77, 82, 94, 111; starting with a, 7

perfectionism, 5, 6, 112

persistence, 128, 134

phylogeny, 55–57

pilot studies, 3–4

playfulness, 101

pointer, 170

pollination, 10

Pomodoro technique, 113

posters, 172, 175–179; checklist, 178–179

pot size, 60

power: analysis, 73–74; blocks, 53; statistical, 44, 45, 52, 53, 191

practice, oral presentations, 165–170

precision, 43

predictors, 71, 89–91

preliminary data. See data, preliminary

principal component analysis (PCA), 90

probability, 43, 61–64

procrastination, 108

professor, major, 6, 9, 16, 109, 116–119, 130, 158, 189

project: question-oriented, 7–9; system-oriented, 7, 9–11

proposals, grant and research: 179–185, 187–189; checklist, 186–187

pseudoreplication, 50–52

pseudorigor, 93

publications, 34, 118, 122; currency, xiii-xiv, 126, 192; process, 154, 157–158

punch line. See take-home message

questions: 1–16, 191; and approaches, 17, 34; applied vs. basic, 2–3; building a story, 13–14, 160; clear, 35–36, 77, 85, 146–147, 191; and experiments, 44, 57, 60, 66, 76, 109; generating, 9–11, 12, 20, 36–37, 100, 101–105; in articles, 143–147; in proposals, 182–184, 188–189; and reading, 106–108; in talks, 159–160, 164–165; on posters, 175–176; posing, 64; yes/no, 66–67, 70; and your goals, 14–16

randomization, 38, 50–53, 191; effect, 54–55; lack of, 22; reassignment, 52

range, 40–41, 46

reading, 5, 106–108, 112, 131, 145; during talks, 163, 166, 168

realism, 19, 21, 45, 59, 87, 191

regression, 70–71, 90; multiple, 54, 70, 89–90; partial, 90; phylogenetically explicit, 57; spatially explicit, 54–55

rejections, 137, 154, 158

replication, 3, 8, 21, 41–46, 191; data sheets, 110; independence, 50–53; non-independence, 53–57; power, 73; pseudo, 50–51; temporal, 45; vs scale, 45–46

requisites, study, 37

research. *See* question, project, *or* proposal

response variables, 84–86

resubmission, 158, 188

results: champagne, 189; communicating, 137, 141; extrapolating/generalizing, 21, 24, 47–48, 74, 81, 153; imprecision, 43; in articles, 137, 141–143, 146, 148–153, 156–157; in talks, 159–161, 168–169, 173–174; modeling, 24, 28; negative, 72–74; organizing, 108, 111, 112; on posters, 176, 178; preliminary, 182; presentation of, 62; pressure to produce, 68, 83; in proposals, 183, 185, 186

review: articles, 106; literature, 112, 138

Rick, world of, 30–31, 36, 59–60, 85, 129, 140, 166

ridicule, 99

risk taking, 99

rules: brainstorming, 102; unwritten, 127, 128, 190–192

saliva, 40

sample size, 43–45, 55, 63, 73, 90, 92, 111; biological significance, 95; scale, 45

sampling: biased, 43–44, 64; non-independence, 54–57; subsampling, 41–42, 50

scale: 76, 81, 94–95, 191; collaborators, 40; effect on worldview, 48; and replication, 45–46; spatial, 21–22, 45, 81; temporal, 21–22, 81

scatterplots, 151, 156

scientific reasoning, 17

scope, 46–47, 81, 83, 121; diagram, 46–47

self: essential, 125–126; social, 125–126

seminars: attending, 108; bullies, 172

sex and babies, 65

significance: biological, 61–62, 95, 150; statistical, 61–65, 73, 95, 150

signposts, in talks, 164–165, 174

sites: field, 8, 83, 109; long-term, 83; matching, 58

skills, 25, 30, 127–128

skimming, 9, 107, 181

slides, 162–168; clarity, 162, 164; dark backgrounds, 170; number of, 167; as reminders, 166

social media, 122

space-for-time, 82–83

speaking: non-native speakers, 142, 157, 163, 168–169; during talks, 163, 166–167, 171

standard deviation (sd), 151

standard error (se), 151, 156

statement: diversity, equity, and inclusion (DEI), 131–132; of purpose, 117; research, 129; teaching, 130–131

statistics, 61–75; Bayesian, 71–72; in papers, 148–150, 151; -speak, 150; tests, 44, 148; value of, 44, 46

storks and human birth rate, 65–66

story, 13–14, 34, 61, 86, 108–109, 139, 141; alternative hypotheses, 68, 111; in papers, 148–153; take-home message, 141; in talks, 158–162
structural equation model, 33, 91–93
subheadings, 149
subsampling. *See* sampling
success, 4–6, 7, 29, 75, 107, 118, 128, 130, 132, 181, 189
summary, 143, 182–183
support: social, 99, 102, 116–119, 131–132; financial, *see* funding
survey. *See* observations
system: finding one, 8–10; native, 8–9

tables: in papers, 149–151; in talks, 161–162
take-home message, 137, 141; in papers, 154; in posters, 175–176; in talks, 159, 163, 169
talks. *See* oral presentations
Tao of ecology, 29–30
teaching: experience, 127; jobs, 129–132; philosophy 130–131
theory, in talks, 163–164
time management: field work, 108–112; reading, 106–108
timetable, grants, 182, 187
title: article, 143; poster, 175–176
treatments, 22, 23, 59, 84, 87, 110; confounded, 50–52; and controls, 23, 41, 49; blocking, 53; haphazard assignment of, 52; interspersion of, 50–53; levels, 23; meaningful, 38–41; random assignment of, 52–53; replication of, 41–48; side effects, 39;
troubleshooting, 109–110, 185
true/false hypotheses, 66–68, 191
truth, 115
turtles, 25

universality, 67, 191
universities, model for, 115–116

vandalism, 8
variability, 59
variance inflation factor (VIF), 90
variation: natural, 10, 12, 19, 20, 40–41, 45, 88–89; and p-values, 63; partitioning, 70; as source of questions, 12, 59; space-for-time, 82–83; standard deviation, 151

weather, 32
Web of Science, 106, 116, 131
White, Tom, 32–33, 93
wooly bear caterpillars, 85
work-life balance, 101, 116
writing: brainstorming, 98; currency, xiv, 192; field notebook, 20; field season, 109–112; habits, 113–114; journal articles, 143–154; non-native English, 142, 157; process, 107–108, 109–112, 137, 138–143; proposals, 138–143, 180, 188; questions, 107, 109; talks, 159–161, 163

yes/no, 67, 70